Data Wrangling

Scrivener Publishing
100 Cummings Center, Suite 541J
Beverly, MA 01915-6106

Publishers at Scrivener
Martin Scrivener (martin@scrivenerpublishing.com)
Phillip Carmical (pcarmical@scrivenerpublishing.com)

Data Wrangling

Concepts, Applications and Tools

Edited by
M. Niranjanamurthy
Kavita Sheoran
Geetika Dhand
and
Prabhjot Kaur

Scrivener Publishing

WILEY

This edition first published 2023 by John Wiley & Sons, Inc., 111 River Street, Hoboken, NJ 07030, USA and Scrivener Publishing LLC, 100 Cummings Center, Suite 541J, Beverly, MA 01915, USA

For more information about Scrivener publications please visit www.scrivenerpublishing.com.

Wiley Global Headquarters
111 River Street, Hoboken, NJ 07030, USA

For details of our global editorial offices, customer services, and more information about Wiley products visit us at www.wiley.com.

Library of Congress Cataloging-in-Publication Data

ISBN 978-1-119-87968-8

Cover images: Color Grid Background | Anatoly Stojko | Dreamstime.com
Data Center Platform | Siarhei Yurchanka | Dreamstime.com
Cover design: Kris Hackerott

Set in size of 11pt and Minion Pro by Manila Typesetting Company, Makati, Philippines

Printed in the USA

10 9 8 7 6 5 4 3 2 1

Contents

1

Basic Principles of Data Wrangling

Akshay Singh*, Surender Singh and Jyotsna Rathee

Department of Information Technology, Maharaja Surajmal Institute of Technology, Janakpuri, New Delhi, India

Abstract

Data wrangling is considered to be a crucial step of data science lifecycle. The quality of data analysis directly depends on the quality of data itself. As the data sources are increasing with a fast pace, it is more than essential to organize the data for analysis. The process of cleaning, structuring, and enriching raw data into the required data format in order to make better judgments in less time is known as data wrangling. It entails the manual conversion and mapping of data from one raw form to another in order to facilitate data consumption and organization. It is also known as data munging, meaning "digestible" data. The iterative process of gathering, filtering, converting, exploring, and integrating data come under the data wrangling pipeline. The foundation of data wrangling is data gathering. The data is extracted, parsed, and scraped before the process of removing unnecessary information from raw data. Data filtering or scrubbing includes removing corrupt and invalid data, thus keeping only the needful data. The data is transformed from unstructured to a bit structured form. Then, the data is converted from one format to another format. To name a few, some common formats are CSV, JSON, XML, SQL, etc. The preanalysis of data is to be done in data exploration step. Some preliminary queries are applied on the data to get the sense of the available data. The hypothesis and statistical analysis can be formed after basic exploration. After exploring the data, the process of integrating data begins in which the smaller pieces of data are added up to form big data. After that, validation rules are applied on data to verify its quality, consistency, and security. In the end, analysts prepare and publish the wrangled data for further analysis. Various platforms available for publishing the wrangled data are GitHub, Kaggle, Data Studio, personal blogs, websites, etc.

***Keywords*:** Data wrangling, big data, data analysis, cleaning, structuring, validating, optimization

**Corresponding author*: akshaysingh@msit.in

M. Niranjanamurthy, Kavita Sheoran, Geetika Dhand, and Prabhjot Kaur (eds.) Data Wrangling: Concepts, Applications and Tools, (1–18) © 2023 Scrivener Publishing LLC

1.1 Introduction

Meaningless raw facts and figures are termed as data which are of no use. Data are analyzed so that it provides certain meaning to raw facts, which is known as information. In current scenario, we have ample amount of data that is increasing many folds day by day which is to be managed and examined for better performance for meaningful analysis of data. To answer such inquiries, we must first wrangle our data into the appropriate format. The most time-consuming part and essential part is wrangling of data [1].

Definition 1—"Data wrangling is the process by which the data required by an application is identified, extracted, cleaned and integrated, to yield a data set that is suitable for exploration and analysis." [2]

Definition 2—"Data wrangling/data munging/data cleaning can be defined as the process of cleaning, organizing, and transforming raw data into the desired format for analysts to use for prompt decision making."

Definition 3—"Data wrangling is defined as an art of data transformation or data preparation." [3]

Definition 4—"Data wrangling term is derived and defined as a process to prepare the data for analysis with data visualization aids that accelerates the faster process." [4]

Definition 5—"Data wrangling is defined as a process of iterative data exploration and transformation that enables analysis." [1]

Although data wrangling is sometimes misunderstood as ETL techniques, these two are totally different with each other. Extract, transform, and load ETL techniques require handiwork from professionals and professionals at different levels of the process. Volume, velocity, variety, and veracity, i.e., 4 V's of big data becomes exorbitant in ETL technology [2].

We can categorize values into two sorts along a temporal dimension in any phase of life where we have to deal with data: near-term value and long-term value. We probably have a long list of questions we want to address with our data in the near future. Some of these inquiries may be ambiguous, such as "Are consumers actually changing toward communicating with us via their mobile devices?" Other, more precise inquiries can include: "When will our clients' interactions largely originate from mobile devices rather than desktops or laptops?" Various research work, different projects, product sale, company's new product to be launched, different businesses etc. can be tackled in less time with more efficiency using data wrangling.

- Aim of Data Wrangling: Data wrangling aims are as follows:
  a) Improves data usage.
  b) Makes data compatible for end users.
  c) Makes analysis of data easy.
  d) Integrates data from different sources, different file formats.
  e) Better audience/customer coverage.
  f) Takes less time to organize raw data.
  g) Clear visualization of data.

In the first section, we demonstrate the workflow framework of all the activities that fit into the process of data wrangling by providing a workflow structure that integrates actions focused on both sorts of values. The key building pieces for the same are introduced: data flow, data wrangling activities, roles, and responsibilities [10]. When commencing on a project that involves data wrangling, we will consider all of these factors at a high level.

The main aim is to ensure that our efforts are constructive rather than redundant or conflicting, as well as within a single project by leveraging formal language as well as processes to boost efficiency and continuity. Effective data wrangling necessitates more than just well-defined workflows and processes.

Another aspect of value to think about is how it will be provided within an organization. Will organizations use the exact values provided to them and analyze the data using some automated tools? Will organizations use the values provided to them in an indirect manner, such as by allowing employees in your company to pursue a different path than the usual?

- Indirect Value: By influencing the decisions of others and motivating process adjustments. In the insurance industry, for example, risk modeling is used.
- Direct Value: By feeding automated processes, data adds value to a company. Consider Netflix's recommendation engine [6].

Data has a long history of providing indirect value. Accounting, insurance risk modeling, medical research experimental design, and intelligence analytics are all based on it. The data used to generate reports and visualizations come under the category of indirect value. This can be accomplished when people read our report or visualization, assimilate the information into their existing world knowledge, and then apply that knowledge to improve their behaviors. The data here has an indirect influence on other people's judgments. The majority of our data's known potential value will be given indirectly in the near future.

Giving data-driven systems decisions for speed, accuracy, or customization provides direct value from data. The most common example is resource distribution and routing that is automated. This resource is primarily money in the field of high-frequency trading and modern finance. Physical goods are routed automatically in some industries, such as Amazon or Flipkart. Hotstar and Netflix, for example, employ automated processes to optimize the distribution of digital content to their customers. For example, antilock brakes in automobiles employ sensor data to channel energy to individual wheels on a smaller scale. Modern testing systems, such as the GRE graduate school admission exam, dynamically order questions based on the tester's progress. A considerable percentage of operational choices is directly handled by data-driven systems in all of these situations, with no human input.

1.2 Data Workflow Structure

In order to derive direct, automated value from our data, we must first derive indirect, human-mediated value. To begin, human monitoring is essential to determine what is "in" our data and whether the data's quality is high enough to be used in direct and automated methods. We cannot anticipate valuable outcomes from sending data into an automated system blindly. To fully comprehend the possibilities of the data, reports must be written and studied. As the potential of the data becomes clearer, automated methods can be built to utilize it directly. This is the logical evolution of information sets: from immediate solutions to identified problems to longer-term analyses of a dataset's fundamental quality and potential applications, and finally to automated data creation systems. The passage of data through three primary data stages:

a) raw,
b) refined,
c) produced,

is at the heart of this progression.

1.3 Raw Data Stage

In the raw data stage, there are three main actions: data input, generic metadata creation, and proprietary metadata creation. As illustrated in

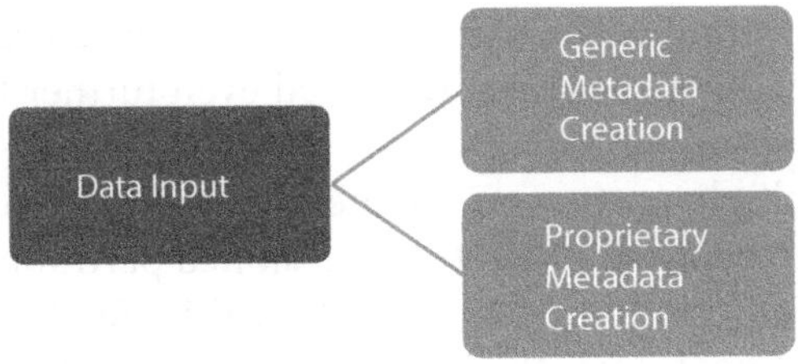

Figure 1.1 Actions in the raw data stage.

Figure 1.1, based on their production, we can classify these actions into two groups. The two ingestion actions are split into two categories, one of which is dedicated to data output. The second group of tasks is metadata production, which is responsible for extracting information and insights from the dataset.

The major purpose of the raw stage is to uncover the data. We ask questions to understand what our data looks like when we examine raw data. Consider the following scenario:

- What are the different types of records in the data?
- How are the fields in the records encoded?
- What is the relationship between the data and our organization, the kind of processes we have, and the other data we already have?

1.3.1 Data Input

The ingestion procedure in traditional enterprise data warehouses includes certain early data transformation processes. The primary goal of these transformations is to transfer inbound components to their standard representations in the data warehouse.

Consider the case when you are ingesting a comma separated file. The data in the CSV file is saved in predetermined locations after it has been modified to fit the warehouse's syntactic criteria. This frequently entails adding additional data to already collected data. In certain cases, appends might be as simple as putting new records to the "end" of a dataset. The add procedure gets more complicated when the incoming data contains both changes to old data and new data. In many of these instances, you will need to ingest fresh data into a separate place, where you can apply more intricate merging criteria during the refined data stage. It is important to highlight, however, that a separate refined data stage will be required throughout the entire spectrum of ingestion infrastructures. This is due to the fact that refined data has been wrangled even further to coincide with anticipated analysis.

Data from multiple partners is frequently ingested into separate datasets, in addition to being stored in time-versioned partitions. The ingestion logic is substantially simplified as a result of this. As the data progresses through the refinement stage, the individual partner data is harmonized to a uniform data format, enabling for quick cross-partner analytics.

1.3.2 Output Actions at Raw Data Stage

In most circumstances, the data you are consuming in first stage is predefined, i.e., what you will obtain and how to use it are known to you. What will when some new data is added to the database by the company? To put it another way, what can be done when the data is unknown in part or in whole? When unknown data is consumed, two additional events are triggered, both of which are linked to metadata production. This process is referred to as "generic metadata creation." A second activity focuses on determining the value of your data based on the qualities of your data. This process is referred to as "custom metadata creation."

Let us go over some fundamentals before we get into the two metadata-generating activities. Records are the building blocks of datasets. Fields are what make up records. People, items, relationships, and events are frequently represented or corresponded to in records. The fields of a record describe the measurable characteristics of an individual, item, connection, or incident. In a dataset of retail transactions, for example, every entry could represent a particular transaction, with fields denoting the purchase's monetary amount, the purchase time, the specific commodities purchased, etc.

In relational database, you are probably familiar with the terms "rows" and "columns." Rows contain records and columns contain fields. Representational consistency is defined by structure, granularity, accuracy, temporality, and scope. As a result, there are also features of a dataset that your wrangling efforts must tune or improve. The data discovery process frequently necessitates inferring and developing specific information linked to the potential value of your data, in addition to basic metadata descriptions.

1.3.3 Structure

The format and encoding of a dataset's records and fields are referred to as the dataset's structure. We can place datasets on a scale based on how homogeneous their records and fields are. The dataset is "rectangular" at one end of the spectrum and can be represented as a table. The table's rows contain records and columns contain fields in this format. You may be dealing with a "jagged" table when the data is inconsistent. A table like this is not completely rectangular any longer. Data formats like XML and JSON can handle data like this with inconsistent values.

Datasets containing a diverse set of records are further along the range. A heterogeneous dataset from a retail firm, for example, can include both customer information and customer transactions. When considering the tabs in a complex Excel spreadsheet, this is a regular occurrence. The majority of analysis and visualization software will need that these various types of records be separated and separate files are formed.

1.3.4 Granularity

A dataset's granularity relates to the different types of things that represents the data. Data entries represent information about a large number of different instances of the same type of item. The roughness and refinement of granularity are often used phrases. This refers to the depth of your dataset's records, or the number of unique entities associated with a single entry, in the context of data. A data with fine granularity might contain an entry indicating one transaction by only one consumer.

You might have a dataset with even finer granularity, with each record representing weekly combined revenue by location. The granularity of the dataset may be coarse or fine, depending on your intended purpose. Assessing the granularity of a dataset is a delicate process that necessitates the use of organizational expertise. These are some examples of granularity-related custom metadata.

1.3.5 Accuracy

The quality of a data is measured by the accuracy. The records used to populate the dataset's fields should be consistent and correct. Consider the case of a customer activities dataset. This collection of records includes information on when clients purchased goods. The record's identification may be erroneous in some cases; for example, a UPC number can have missing digits or it can be expired. Any analysis of the dataset would be limited by inaccuracies, of course. Spelling mistakes, unavailability of the variables, numerical floating value mistakes, are all examples of common inaccuracies.

Some values can appear more frequently and some can appear less frequently in a database. This condition is called frequency outliers which can also be assessed with accuracy. Because such assessments are based on the knowledge of an individual organization and making frequency assessments is essentially a custom metadata matter.

1.3.6 Temporality

A record present in the table is a snapshot of a commodity at a specific point of time. As a result, even if a dataset had a consistent representation at the development phase and later some changes may cause it to become inaccurate or inconsistent. You could, for example, utilize a dataset of consumer actions to figure out how many goods people own. However, some of these things may be returned weeks or months after the initial transaction. The initial dataset is not the accurate depiction of objects purchased by a customer, despite being an exact record of the original sales transaction.

The time-sensitive character of representations, and thus datasets, is a crucial consideration that should be mentioned explicitly. Even if time is not clearly recorded, then also it is very crucial to know the influence of time on the data.

1.3.7 Scope

A dataset's scope has two major aspects. The number of distinct properties represented in a dataset is the first dimension. For example, we might know when a customer action occurred and some details about it. The second dimension is population coverage by attribute. Let us start with the number of distinct attributes in a dataset before moving on to the importance of scope. In most datasets, each individual attribute is represented by a separate field. There exists a variety of fields in a dataset with broad scope and in case of datasets with narrow scope, there exists a few fields.

The scope of a dataset can be expanded by including extra field attributes. Depending on your analytics methodology, the level of detail necessary may vary. Some procedures, such as deep learning, demand for keeping a large number of redundant attributes and using statistical methods to reduce them to a smaller number. Other approaches work effectively with a small number of qualities. It is critical to recognize the systematic biasness in a dataset since any analytical inferences generated from the biased dataset would be incorrect. Drug trial datasets are usually detailed to the patient level. If, on the other hand, the scope of the dataset has been deliberately changed by tampering the records of patients due to their death during trial or due to abnormalities shown by the machine, the analysis of the used medical dataset is shown misrepresented.

1.4 Refined Stage

We can next modify the data for some better analysis by deleting the parts of the data which have not used, by rearranging elements with bad structure, and building linkages across numerous datasets once we have a good knowledge of it. The next significant part is to refine the data and execute a variety of analysis after ingesting the raw data and thoroughly comprehending its metadata components. The refined stage, Figure 1.2, is defined by three main activities: data design and preparation, ad hoc reporting analysis, and exploratory modelling and forecasting. The first group focuses on the production of refined data that can be used in a variety of studies right away. The second group is responsible for delivering data-driven insights and information.

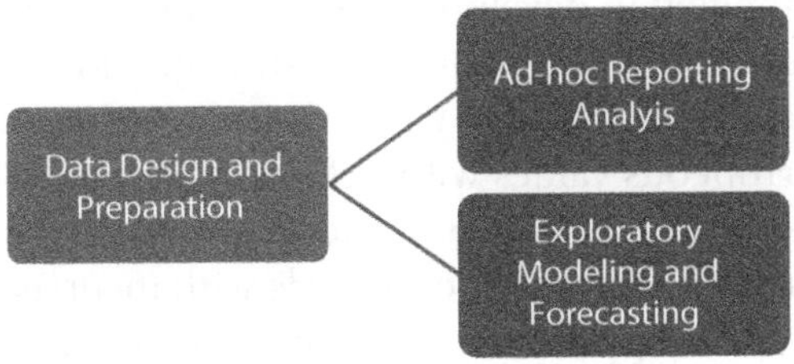

Figure 1.2 Actions in the refined stage.

1.4.1 Data Design and Preparation

The main purpose of creating and developing the refined data is to analyze the data in a better manner. Insights and trends discovered from a first set of studies are likely to stimulate other studies. In the refined data stage, we can iterate between operations, and we do so frequently.

Ingestion of raw data includes minimum data transformation—just enough to comply with the data storage system's syntactic limitations. Designing and creating "refined" data, on the other hand, frequently necessitates a large change. We should resolve any concerns with the dataset's structure, granularity, correctness, timing, or scope that you noticed earlier during the refined data stage.

1.4.2 Structure Issues

Most visualization and analysis tools are designed to work with tabular data, which means that each record has similar fields in the given order. Converting data into tabular representation can necessitate considerable adjustments depending on the structure of the underlying data.

1.4.3 Granularity Issues

It is best to create refined datasets with the highest granularity resolution of records you want to assess. We should figure out what distinguishes the customers that have larger purchases from the rest of customers: Is it true that they are spending more money on more expensive items? Do you have a greater quantity of stuff than the average person? For answering such questions, keeping a version of the dataset at this resolution may be helpful. Keeping numerous copies of the same data with different levels of granularity can make subsequent analysis based on groups of records easier.

1.4.4 Accuracy Issues

Another important goal in developing and refining databases is to address recognized accuracy difficulties. The main strategies for dealing with accuracy issues by removing records with incorrect values and Imputation, which replaces erroneous values with default or estimated values.

In certain cases, eliminating impacted records is the best course of action, particularly when number of records with incorrect values is minimal and unlikely to be significant. In many circumstances, removing these data will have little influence on the outcomes. In other cases, addressing inconsistencies in data, such as recalculating a client's age using their date of birth and current date, may be the best option (or the dates of the events you want to analyze).

Making an explicit reference to time is often the most effective technique to resolve conflicting or incorrect data fields in your refined data. Consider the case of a client database with several addresses. Perhaps each address is (or was) correct, indicating a person's several residences during her life. By giving date ranges to the addresses, the inconsistencies may be rectified. A transaction amount that defies current business logic may have happened before the logic was implemented, in which case the transaction should be preserved in the dataset to ensure historical analysis integrity.

In general, the most usable understanding of "time" involves a great deal of care. For example, there may be a time when an activity happened and a time when it was acknowledged. When it comes to financial transactions, this is especially true. In certain cases, rather than a timestamp, an abstract version number is preferable. When documenting data generated by software, for example, it may be more important to record the software version rather than the time it was launched. Similarly, knowing the version of a data file that was inspected rather than the time that the analysis was run may be more relevant in scientific study. In general, the optimum time or version to employ depends on the study's characteristics; as a result, it is important to keep a record of all timestamps and version numbers.

1.4.5 Scope Issues

Taking a step back from individual record field values, it is also important to make sure your refined datasets include the full collection of records and record fields. Assume that your client data is split into many datasets (one containing contact information, another including transaction summaries, and so on), but that the bulk of your research incorporate all of these variables. You could wish to create a totally blended dataset with all of these fields to make your analysis easier.

Ensure that the population coverage in your altered datasets is understood, since this is likely the most important scope-related issue. This means that a dataset should explain the relationship between the collection of items represented by the dataset's records (people, objects, and so on) and the greater population of those things in an acceptable manner (for example, all people and all objects) [6].

1.4.6 Output Actions at Refined Stage

Finally, we will go through the two primary analytical operations of the refined data stage: ad hoc reporting analyses and exploratory modelling and forecasting. The most critical step in using your data to answer specific questions is reporting. Dash boarding and business intelligence analytics are two separate sorts of reporting.

The majority of these studies are retrospective, which means they depend on historical data to answer questions about the past or present. The answer to such queries might be as simple as a single figure or statistic, or as complicated as a whole report with further discussion and explanation of the findings.

Because of the nature of the first question, an automated system capable of consuming the data and taking quick action is doubtful. The consequences, on the other hand, will be of indirect value since they will inform and affect others. Perhaps sales grew faster than expected, or perhaps transactions from a single product line or retail region fell short of expectations. If the aberration was wholly unexpected, it must be assessed from several perspectives. Is there an issue with data quality or reporting? If the data is authentic (i.e., the anomaly represents a change in the world, not just in the dataset's portrayal of the world), can an anomaly be limited to a subpopulation? What additional alterations have you seen as a result of the anomaly? Is there a common root change to which all of these changes are linked through causal dependencies?

Modeling and forecasting analyses are often prospective, as opposed to ad hoc assessments, which are mostly retrospective. "Based on what we've observed in the past, what do we expect to happen?" these studies ask. Forecasting aims to anticipate future events such as total sales in the next quarter, customer turnover percentages next month, and the likelihood of each client renewing their contracts, among other things. These forecasts are usually based on models that show how other measurable elements of your dataset impact and relate to the objective prediction. The underlying model itself, rather than a forecast, is the most helpful conclusion for some analyses. Modeling is, in most cases, an attempt to comprehend the important factors that drive the behavior that you are interested in.

1.5 Produced Stage

After you have polished your data and begun to derive useful insights from it, you will naturally begin to distinguish between analyses that need to be repeated on a regular basis and those that can be completed once. Experimenting and prototyping (which is the focus of activities in the refined data stage) is one thing; wrapping those early outputs in a dependable, maintainable framework that can automatically direct people and resources is quite another. This places us in the data-gathering stage. Following a good set of early discoveries, popular comments include, "We should watch that statistic all the time," and "We can use those forecasts to speed up shipping of specific orders." Each of these statements has a solution using "production systems," which are systems that are largely automated and have a well-defined level of robustness. At the absolute least, creating production data needs further modification of your model. The action steps included in the produced stage are shown in Figure 1.3.

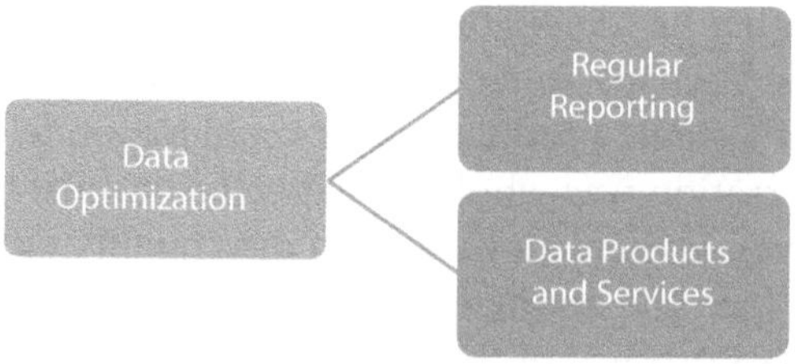

Figure 1.3 Actions in the produced stage.

1.5.1 Data Optimization

Data refinement is comparable to data optimization. The optimum form of your data is optimized data, which is meant to make any further downstream effort to use the data as simple as feasible.

There are also specifications for the processing and storage resources that will be used on a regular basis to work with the data. The shape of the data, as well as how it is made available to the production system, will frequently be influenced by these constraints. To put it another way, while the goal of data refinement is to enable as many studies as possible as quickly as possible, the goal of data optimization is to facilitate a relatively small number of analysis as consistently and effectively as possible.

1.5.2 Output Actions at Produced Stage

More than merely plugging the data into the report production logic or the service providing logic is required for creating regular reports and data-driven products and services. Monitoring the flow of data and ensuring that the required structural, temporal, scope, and accuracy criteria are met over time is a substantial source of additional effort. Because data is flowing via these systems, new (or updated) data will be processed on a regular basis. New data will ultimately differ from its historical counterparts (maybe you have updated customer interaction events or sales data from the previous week). The border around allowable variation is defined by structural, temporal, scope, and accuracy constraints (e.g., minimum and maximum sales amounts or coordination between record variables like billing address and transaction currency). The reporting and product/service logic must handle the variation within the restrictions [6].

This differs from exploratory analytics, which might use reasoning specific to the dataset being studied for speed or simplicity. The reasoning must be generalized for production reporting and products/services. Of course, you may narrow the allowable variations boundary to eliminate duplicate records and missing subsets of records. If that is the case, the logic for detecting and correcting these inconsistencies will most likely reside in the data optimization process.

Let us take a step back and look at the fundamentals of data use to assist motivate the organizational changes. Production uses, such as automated reports or data-driven services and products, will be the most valuable uses of your data. However, hundreds, if not thousands, of exploratory, ad hoc analyses are required for every production usage of your data. In other words, there is an effort funnel that starts with exploratory analytics

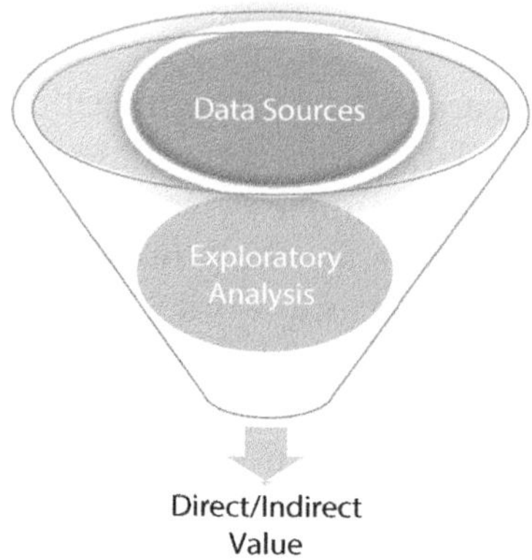

Figure 1.4 Data value funnel.

and leads to direct, production value. Your conversation rate will not be 100%, as it is with any funnel. In order to identify a very limited number of meaningful applications of your data, you will need as many individuals as possible to explore it and derive insights. A vast number of raw data sources and exploratory analysis are necessary to develop a single useful application of your data, as shown in Figure 1.4.

When it comes to extracting production value from your data, there are two key considerations. For starters, data might provide you and your firm with useless information. These insights may not be actionable, or their potential impact may be too little to warrant a change in current practices. Empowering the people who know your business priorities to analyze your data is a smart strategy for mitigating this risk. Second, you should maximize the efficiency of your exploratory analytics activities. Now we are back to data manipulation. The more data you can wrangle in a shorter amount of time, the more data explorations you can do and the more analyses you can put into production.

1.6 Steps of Data Wrangling

We have six steps, as shown in Figure 1.5, for data wrangling to convert raw data to usable data.

a) Discovering data—Data that is to be used is to be understood carefully and is collected from different sources in different range of formats and sizes to find patterns and trends. Data collected from different sources and in different format are well acknowledged [7].

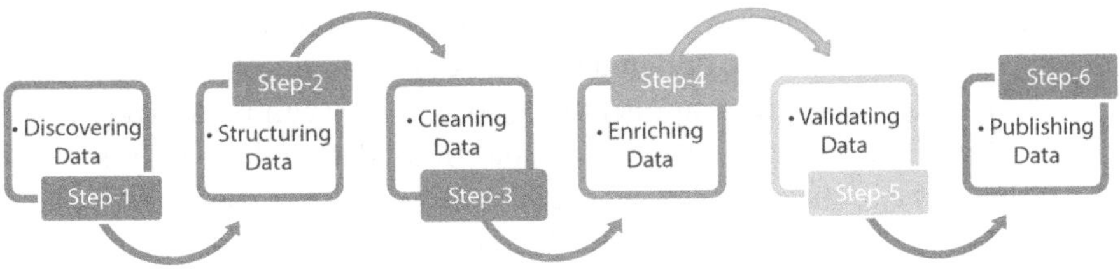

Figure 1.5 Steps for data wrangling process.

b) Structuring data—Data is in unstructured format or disorganized while collecting data from different sources, so data is organized and structured according to Analytical Model of the business or according to requirement. Relevant information is extracted from data and is organized in structured format. For Example certain columns should be added and certain columns in the data should be removed according to our requirement.

c) Cleaning data—Cleaning data means to clean data so that it is optimum for analysis [8]. As certain outliers are always present in data which reduces analysis consequences. This step includes removing outliers from dataset changes null or empty data with standardized values, removes structural errors [5].

d) Enriching data—The data must be enriched after it has been cleaned, which is done in the enrichment process. The goal is to enrich existing data by adding more data from either internal or external data sources, or by generating new columns from existing data using calculation methods, such as folding probability measurements or transforming a time stamp into a day of the week to improve accuracy of analysis [8].

e) Validating data—In validation step we check quality, accuracy, consistency, security and authenticity of data. The validation process will either uncover any data quality issues or certify that an appropriate transformation has been performed. Validations should be carried out on a number of different dimensions or rules. In any case, it is a good idea to double-check that attribute or field values are proper and meet the syntactic and distribution criteria. For example, instead of 1/0 or [True, False], a Boolean field should be coded as true or false.

f) Publishing data—This is the final publication stage, which addresses how the updated data are delivered to subject analysts and for which applications, so that they can be utilized for other purposes afterward. Analysis of data is done in this step i.e. data is placed where it is accessed and used. Data are placed in a new architecture or database. Final output received is of high quality and more accurate which brings new insights to business. The process of preparing and transferring data wrangling output for use in downstream or future projects, such as loading into specific analysis software or documenting and preserving the transformation logic, is referred to as publishing. When the input data is properly formatted, several analytic tools operate substantially faster. Good data wrangler software understands this and formats the processed data in such a way that the target system can make the most of it. It makes sense to reproduce a project's data wrangling stages and methods for usage on other databases in many circumstances.

1.7 Do's for Data Wrangling

Things to be kept in mind in data wrangling are as follows:

a) Nature of Audience—Nature of audience is to kept in mind before starting data wrangling process.
b) Right data—Right data should be picked so that analysis process is more accurate and of high quality.
c) Understanding of data is a must to wrangle data.
d) Reevaluation of work should be done to find flaws in the process.

1.8 Tools for Data Wrangling

Different tools used for data wrangling process that you will study in this book in detail are as follows [9]:

- MS Excel
- Python and R
- KNIME

- OpenRefine
- Excel Spreadsheets
- Tabula
- PythonPandas
- CSVKit
- Plotly
- Purrr
- Dplyr
- JSOnline
- Splitstackshape

The foundation of data wrangling is data gathering. The data is extracted, parsed and scraped before the process of removing unnecessary information from raw data. Data filtering or scrubbing includes removing corrupt and invalid data, thus keeping only the needful data. The data are transformed from unstructured to a bit structured form. Then, the data is converted from one format to another format. To name a few, some common formats are CSV, JSON, XML, SQL, etc. The preanalysis of data is to be done in data exploration step. Some preliminary queries are applied on the data to get the sense of the available data. The hypothesis and statistical analysis can be formed after basic exploration. After exploring the data, the process of integrating data begins in which the smaller pieces of data are added up to form big data. After that, validation rules are applied on data to verify its quality, consistency and security. In the end, analysts prepare and publish the wrangled data for further analysis.

References

1. Kandel, S., Heer, J., Plaisant, C., Kennedy, J., Van Ham, F., Riche, N.H., Weaver, C., Lee, B., Brodbeck, D., Buono, P., Research directions in data wrangling: Visualizations and transformations for usable and credible data. *Inf. Vis.*, 10, 4, 271–288, 2011.
2. Furche, T., Gottlob, G., Libkin, L., Orsi, G., Paton, N.W., Data wrangling for big data: Challenges and opportunities, in: *EDBT*, vol. 16, pp. 473–478, March 2016.
3. Patil, M.M. and Hiremath, B.N., A systematic study of data wrangling. *Int. J. Inf. Technol. Comput. Sci.*, 1, 32–39, 2018.
4. Cline, D., Yueh, S., Chapman, B., Stankov, B., Gasiewski, A., Masters, D., Elder, K., Kelly, R., Painter, T.H., Miller, S., Katzberg, S., NASA cold land processes experiment (CLPX 2002/03): Airborne remote sensing. *J. Hydrometeorol.*, 10, 1, 338–346, 2009.

5. Dasu, T. and Johnson, T., *Exploratory Data Mining and Data Cleaning*, vol. 479, John Wiley & Sons, Hoboken, New Jersey, United States, 2003.
6. Rattenbury, T., Hellerstein, J.M., Heer, J., Kandel, S., Carreras, C., *Principles of Data Wrangling: Practical Techniques for Data Preparation*, O'Reilly Media, Inc., Sebastopol, California, 2017.
7. Kim, W., Choi, B.J., Hong, E.K., Kim, S.K., Lee, D., A taxonomy of dirty data. *Data Min. Knowl. Discovery*, 7, 1, 81–99, 2003.
8. Azeroual, O., Data wrangling in database systems: Purging of dirty data. *Data*, 5, 2, 50, 2020.
9. Kazil, J. and Jarmul, K., *Data Wrangling with Python: Tips and Tools to Make Your Life Easier*, O'Reilly Media, Inc., Sebastopol, California, 2016.
10. Endel, F. and Piringer, H., Data wrangling: Making data useful again. *IFAC-PapersOnLine*, 48, 1, 111–112, 2015.

2

Skills and Responsibilities of Data Wrangler

Prabhjot Kaur, Anupama Kaushik and Aditya Kapoor*

Department of Information Technology, Maharaja Surajmal Institute of Technology, Janak Puri, New Delhi, India

Abstract

The following chapter will draw emphasis on the right skill set that must be possessed by the administrators to be able to handle the data and draw interpretations from it. Technical skill set includes knowledge of statistical languages, such as R, Python, and SQL. Data administrators also use tools like Excel, PoweBI, Tableau for data visualization. The chapter aims to draw emphasis on the requirement of much needed soft skills, which provide them an edge over easy management of not just the data but also human resources available to them. Soft skills include effective communication between the clients and team to yield the desired results. Presentation skills are certainly crucial for a data engineer, so as to be able to effectively communicate what the data has to express. It is an ideal duty of a data engineer to make the data speak. The effectiveness of a data engineer in their tasks comes when the data speaks for them. The chapter also deals with the responsibilities as a data administrator. An individual who is well aware of the responsibilities can put their skill set and resources to the right use and add on to productivity of his team thus yielding better results. Here we will go through responsibilities like data extraction, data transformation, security, data authentication, data backup, and security and performance monitoring. A well aware administrator plays a crucial role in not just handling the data but the human resource assigned to them. Here, we also look to make readers aware of the consequences of mishandling the data. A data engineer must be aware of the consequences of data mismanagement and how to effectively handle the issues that occurred. At the end, the chapter is concluded with discussion of two case studies of the two companies UBER and PepsiCo and how effective data handling helped them get better results.

**Corresponding author*: 2000aditya28@gmail

M. Niranjanamurthy, Kavita Sheoran, Geetika Dhand, and Prabhjot Kaur (eds.) Data Wrangling: Concepts, Applications and Tools, (19–52) © 2023 Scrivener Publishing LLC

Keywords: Data administrator, data handling, soft skills, responsibilities, data security, data breaching

2.1 Introduction

In a corporate setup, someone who is responsible for processing huge amounts of data in a convenient data model is known as a data administrator [1]. Their role is primarily figuring out which data is more relevant to be stored in the given database that they are working on. This job profile is basically less technical and requires more business acumen with only a little bit of technical knowledge. Data administrators are commonly known as data analysts. The main crux of their job responsibility is that they are responsible for overall management of data, and it is associated resources in a company.

However, at times, the task of the data administrators is being confused with the database administrator (DBA). A database administrator is specifically a programmer who creates, updates and maintains a database. Database administration is DBMS specific. The role of a data administrator is more technical and they are someone who is hired to work on a database and optimize it for high performance. Alongside they are also responsible for integrating a database into an application. The major skills required for this role are troubleshooting, logical mindset and keen desire to learn along with the changes in the database. The role of a database administrator is highly varied and involves multiple responsibilities. A database administrator's work revolves around database design, security, backup, recovery, performance tuning, etc.

Data scientist is a professional responsible for working on extremely large datasets whereby they inculcate much needed programming and hard skills like machine learning, deep learning, statistics, probability and predictive modelling [2]. A data scientist is the most demanding job of the decade. A data scientist role involves studying data collected, cleaning it, drawing visualizations and predictions from the already collected data and henceforth predicting the further trends in it. As a part of the skill set, a data scientist must have strong command over python, SQL, and ability to code deep neural networks.

The data scientists as professionals are in huge demand since the era of data exploration has begun. As companies are looking forward to extracting only the needed information from big data, huge volumes of structured or unstructured and semistructured data so as to find useful interpretations which will in turn help in increasing the company's profits to great extent. Data scientist basically decent on the creative insights drawn from big data, or information collected via processes, like data mining.

2.2 Role as an Administrator (Data and Database)

Data administrators are supposed to render some help to the various other departments like the ones dealing with marketing, sales, finance, and operations divisions by providing them with the data that they need so that all the information concerning product, customer and vendor is accurate, complete and current. As a data administrator, they will basically implement and execute the data mining projects and further create reports using investigative, organizational and analytical skills, to give and have some sales insights. This way, they also get knowledge about different and crucial factors like purchasing opportunity and trends that follow. The job profile is not just restricted to it but it also includes making needed changes or updates in the database of the company and their website. Their tasks include reporting, performing data analysis, forecasting, market assessments and carrying out various other research activities that play an important role in decision making. They play with data according to the need and requirements of the management. A data administrator is also responsible for updating the data of the vendors and products in the company's database. Not only this but a DBA is also responsible for installing the database softwares [3]. They are also supposed to configure the softwares and according to the requirements they need to upgrade them if needed. Some of the database tools include oracle, MySQL and Microsoft SQL. It is sole responsibility of the DBA to decide how to install these softwares and configure them accordingly [4]. A DBA basically acts as an advisor to the team of database managers and app developers in the company as well. A DBA is expected to be well acquainted with technologies and products like SQL DBA, APIs like JDBC, SQLJ, ODBC, REST, etc., interfacers, encoders, and frameworks, like NET, Java EE, and more.

If we become more specific in terms of roles then a person who works specifically in the warehousing domain is known as data warehouse administrator. As a warehouse administrator, they would specifically need expertise in the domains like:

- Query tools, BI (Business intelligence) applications, etc.;
- OLTP data warehousing;
- Data warehousing specialized designs;
- ETL skills;
- Knowledge of data warehousing technology, various schemas for designs, etc.

Cloud DBA. In today's world of ever growing data, all the companies and organizations are moving over to the cloud. This has increased the demand of Cloud DBA [5]. The work profile is more or less similar to that of a DBA. It is just that they have switched over to cloud platforms for working. The DBA must have some level of proficiency especially in the implementation on Microsoft Azure, AWS, etc. They should be aware of what is involved in tasks related to security and backup functions on cloud, cloud database implementations. They also look into factors, like latency, cost management, and fault tolerance.

2.3 Skills Required

2.3.1 Technical Skills

It is important to be technically sound and possess some basic skill set to play with data. Here, we describe the need to have skills to work with data and draw inference from the data. The following skills will facilitate your learning and ability to draw inference from the data. The following programming languages and tools pave the way to import and study datasets containing millions of entries in a simplified way.

2.3.1.1 Python

A large amount of the coding population has a strong affinity toward python as a programming language. The first time python was used in 1991 and thereafter it has made a strong user base. It has become one of the most widely used languages credits to its easy understandability. Because of its features like being easily interpreted, and various other historical and cultural reason, Pythonists have come up as a larger community of people in the domain of data analysis and scientific computing [6]. Knowing Python programming has become one of the most basic and crucial tasks to be able to enter the field of data science, machine learning and general software development. But at the same time, due to the presence of other languages, like R, MATLAB, SAS, it certainly draws a lot of comparisons too.

Off late, Python has undoubtedly become an obvious choice because of its widely used libraries like Pandas and scikit-learn. Python is also being used for building data applications, given that it is widely acceptable for software engineering practices. Here we will ponder on a few libraries widely used for data analysis:

a) NumPy: Numerical Python aka NumPy, is a crucial library for numerical computing in Python. It provides the much-needed support required to work on the numerical data and specifically for data analysis.

NumPy contains, among other things:

- It has some crucial functions which make it possible to perform elementwise computations or do some mathematical computations between arrays.
- It also has tools for working with datasets in the form of arrays to the disks.
- It helps us to do various operations related to linear algebra, Fourier transform or random number generation for that very matter.
- Also NumPy facilitates array processing in Python, thus this is one of the most important use of NumPy library in python. It is used for data analysis, whereby it helps to put the data in the form of containers to be passed between the algorithms and libraries.

For numerical data, NumPy arrays have been found to be more efficient in tasks like storage of data and its manipulation in comparison to any other data structures in python.

b) Pandas: The pandas name is derived from panel data. It is basically a term specifically used to describe a multidimensional dataset that is also structured and plays a vital role in Python data analysis itself. It is due to the presence of libraries, like Pandas, which facilitate working with structured data much efficiently and expressively due to the presence of high-level data structures and functions. They have enabled a powerful and much efficient data analysis environment in Python.

The primary object in pandas that is most commonly used is data frame. A data frame is tabular in nature, i.e., column oriented. This data structure has both row and column labels. The series is a 1-D labeled array object. Pandas library perfectly blends the spreadsheets and relational databases (such as SQL) along with high-performance, array-computing ideas of NumPy. Not only this but it also provides an indexing functionality to easily manipulate arrays by reshape, slice and dice, perform aggregations, and select subsets of data. Since data manipulation, preliminaries preparation, and cleaning is such an important skill in data analysis, knowing pandas is one of the primary tasks. Some advantages of Pandas are

- Data structures with labeled axes—this basically acts as a facilitator to prevent common errors to come up that might arise due to misaligned data and at the same time this also helps in working with differently indexed data that might have originated from different sources.
- It also comes with a functionality of integrated time series.
- They help us to undergo various arithmetic operations and reductions that specifically help in reductions that preserve the metadata.
- It is also highly flexible in handling the missing values in the data.

Pandas specifically features deep time series functions which are primarily used by business processes whereby time-indexed data is generated. That is the main reason why main features found in andas are either part of R programming language or is provided by some additional packages.

c)Matplotlib: Matplotlib is one of the most popular Python libraries for producing data visualizations. It facilitates visualization tasks by creating plots and graphs. The plots created using Matplotlib are suitable for publication. Matplotlib's integration with the rest of the ecosystem makes it the most widely used.

The IPython shell and Jupyter notebooks play a great role in data exploration and visualization. The Jupyter notebook system also allows you to author content in Markdown and HTML, providing a way to create documents containing both code and text. IPython is usually used in the majority of Python work, such as running, debugging, and testing code.

d) SciPy: This library is basically a group of packages, which play a significant role in dealing with problems related to scientific computing. Some of these are mentioned here:

scipy.integrate: It is used for tasks like numerical integrations and solving the complex differential equations.

scipy.linalg: This is basically used for solving linear algebra and plays a crucial role in matrix decompositions. They have more than the once provided in numpy.linalg.

scipy.optimize: This function is used as a function optimizer and root finding algorithm.

scipy.signal: This provides us the functionality of signal processing.

scipy.sparse: This helps us to solve sparse matrices or sparse linear systems.

scipy.stats: This is basically used for continuous and discrete probability distribution. Also, this is used for undergoing various statistical tests and a lot more descriptive mathematical computations performed by both Numpy and SciPy libraries. Further enhancement, the sophisticated and scientific computations were easier.

e) Scikit-learn: This library has become one of the most important general purpose Machine learning toolkits for pythonistas. It has various submodules for various classification, regression, clustering, and dimensionality reduction algorithms. It helps in model selection and at the same helps in preprocessing. Various preprocessing tasks that it facilitates include feature selection, normalization.

Along with pandas, IPython, scikit-learn has a significant role in making python one of the most important data science programming languages.

In comparison to scikit-learn, statsmodel also has algorithms which help in classical statistics and econometrics. They include submodules, such as regression models, analysis of variance (ANOVA), time series analysis, and nonparametric methods.

2.3.1.2 R Programming Language [7]

R is an extremely flexible statistics programming language and environment that is most importantly Open Source and freely available for almost all operating systems. R has recently experienced an "explosive growth in use and in user contributed software."

R has ample users and has up-to-date statistical methods for analysis. The flexibility of R is unmatched by any other statistics programming language, as its object-oriented programming language allows for the performance of customized procedures by creating functions that help in automation of most commonly performed tasks.

Currently, R is maintained by the R Core Development Team R being an open source can be improvised by the contributions of users from throughout the world. It just has a base system with an option of adding packages as per needs of the users for a variety of techniques.

It is advantageous to use R as a programming language in comparison to other languages because of its philosophy. In R, statistical analysis is done in a series of steps, and its immediate results are stored in objects, these objects are further interrogated for the information of interest. R can be used in integration with other commonly used statistical programs, such as Excel, SPSS, and SAS.R uses vectorized arithmetic, which implies that most equations are implemented in R as they are written, both for scalar and matrix algebra. To obtain the summary statistics for a matrix instead of a vector, functions can be used in a similar fashion.

R as a programming language for data analysis can successfully be used to create scatterplots, matrix plots, histogram, QQ plot, etc. It is also used for multiple regression analysis. It can effectively be used to make interaction plots.

2.3.1.3 SQL [8]

SQL as a programming has revolutionized how the large volumes of data is being perceived by people and how we work on it. Ample SQL queries play a vital role in small analytics practices. SELECT query can be coupled with function or clauses like MIN, MAX, SUM, COUNT, AVG, GROUP BY, HAVING etc on the very large datasets.

All SQL databases, be it commercial/relational/open-source can be used for any type of processing. Big analytics primarily denote regression or data mining practices. They also cover machine learning or other types of complex processing under them. SQL also helps in extraction of data from various sources using SQL queries. The sophisticated analysis require some good packages like SPSS, R, SAS, and some hands-on proficiency in coding.

Usually, statistical packages load their data to be processed using one or more from the following solutions:

- The data can be directly imported from external files where this data is located. This data can be in the form of Excel, CSV or Text Files.
- They also help in saving the intermediate results from the data sources. These data sources can be databases or excel sheets. These are then saved in common format files and then these files are further imported into various packages. Some commonly used interchanging formats are XML, CSV, and JSON.

In recent times, it has been observed that there are ample options available for data imports. Google Analytics being such a service that is becoming quite known among the data analytics community lately. This helps in importing data from the web servers log simply by using user-defined or some standard ETL procedures. It has been found that NoSQL systems have an edge over this particular domain and a significant presence.

In addition to directly importing via ODBC/JDBC connections, at times it is even possible to undergo a database query in a database server with the help of the statistical package directly. For example R users can query SQLLite databases along with directly having the results from the tables into R workspace.

Basically SQL is used to extract records from the databases that are basically very very huge. They also use relational databases to do the needful. The SELECT statement of SQL has some major powerful clauses for filtering records. It helps in grouping them or doing complex computations. The SQL as a programming language has attained center stage due to high-level syntax, which primarily does not require any core coding for most of the queries. It also implements queries from one platform to another like from all database management systems, from desktop to open source and to the commercial ones.

Not only this but the result of SQL queries can also be saved/stored inside the databases and can easily be exported from DBMS to any of the targets or formats as well, i.e., Excel/CSV, Text File, HTML.

SQL wide adaptability and easy understandability and its relation with relational databases and more NoSQL datastores implement SQL-like query languages. This makes many data analysis and data science tasks accessible to non-programmers.

2.3.1.4 MATLAB

A programming language and multi-paradigm numerical computing environment, MATLAB is the final step in advanced data plotting, manipulation, and organization. It is great for companies interested in big data and powerful in machine learning. Machine learning is widely popular in data science right now as a branch of artificial intelligence, and having a good grasp of its models can put you ahead.

2.3.1.5 Scala [9]

Scala is a high level language that combines functional and object oriented programming with high performance runtimes. Spark is typically used in most cases when dealing with big data. Since Spark was built using Scala, it makes sense that learning it will be a great tool for any data scientist. Scala is a powerful language that can leverage many of the same functions as Python, such as building machine learning models. Scala is a great tool to have in our arsenal as data scientists. We can use it working with data and building machine learning models. SCALA has gained much needed center stage due to SPARK being coded in scala and SPARK also being widely used.

2.3.1.6 EXCEL

The ability to analyze data is a powerful skill set that helps you make better decisions related to data and enhance your understanding of that particular dataset. Microsoft Excel is one of the top tools for data analysis and the built-in pivot tables are arguably the most popular analytic tool. In MS Excel there are a lot more features than just using it for SUM and COUNT. Big companies still make use of excel efficiently to transform huge data into readable forms so as to have clear insights of the same. Functions such as CONCATENATE, VLOOKUP and AVERAGEIF(S) are another set of important functions used in industry to facilitate analysis. Data analysis makes it easy for us to draw useful insights from data and thereafter help us to take important decisions on the basis of insights. Excel helps us to explore the dataset at the same time it helps in cleaning data. VLOOKUP is one of the crucial functions that is basically used in excel to add/merge data from one table to another. Effective use of excel but businesses has led them to new heights and growths.

2.3.1.7 Tableau [10]

In the world of visualizations, Tableau occupies the leader post. Not just being user friendly and effective in drawing visualizations it does not lag behind in creating graphs like pivot table graphs in Excel. Not just restricted to that Tableau has the ability to handle a lot more data and is quite fast in providing good amount of calculations.

- Here users are able to create visuals quite fast and can easily switch between different models so as to compare them. This way they can further implement the best ones.
- Tableau has an ability to manage a lot of data.
- Tableau has a much simplified user interface which further allows them to customize the view.
- Tableau has an added advantage of compiling data from multiple data sources.
- Tableau has an ability to hold multiple visualizations without crashing.

The interactive dashboards created in Tableau help us to create visualizations in an effective way as they can be operated on multiple devices like laptop, tablet and mobile. Drop and drag ability of Tableau is an added advantage. Not only this tableau is highly mobile friendly. The interactive dashboards are streamlined in a way that they can be used on mobile devices. It even helps us to run the R models, and at the same time import results into Tableau with much convenience. Its ability of integration with R is an added advantage and helps to build practical models. This integration amplifies data along with providing visual analytics. This process requires less effort.

Tableau can be used by businesses to make multiple charts so as to get meaningful insights. Tableau facilitates finding quick patterns in the data that can be analyzed with the help of R. This software further helps us to fetch the unseen patterns in the big data and the visualizations drawn in Tableau can be used to integrate on the websites.

Tableau has some inbuilt features which help the users in understanding the patterns behind the data and find the reasons behind the correlations and trends. Using tableau basically enhances the user's perspective to look at the things from multiple views and scenarios and this way users can publish data sources separately.

2.3.1.8 Power BI [11]

Main goal as a data analyst is to arrange the insights of our data in such a way that everybody who sees them is able to understand their implications and acts on them accordingly. Power BI is a cloud-based business analytics service from Microsoft that enables anyone to visualize and analyze data, with better speed and efficiency. It is a powerful as well as a flexible tool for connecting with and analyzing a wide variety of data. Many businesses even consider it indispensable for data-science-related work. Power BI's ease of use comes from the fact that it has a drag and drop interface. This feature helps to perform tasks like sorting, comparing and analyzing, very easily and fast. Power BI is also compatible with multiple sources, including Excel, SQL Server, and cloud-based data repositories which makes it an excellent choice for data scientists (Figure 2.1). It gives the ability to analyze and explore data on-premise as well as in the cloud. Power BI provides the ability to collaborate and share customized dashboards and interactive reports across colleagues and organizations, easily and securely.

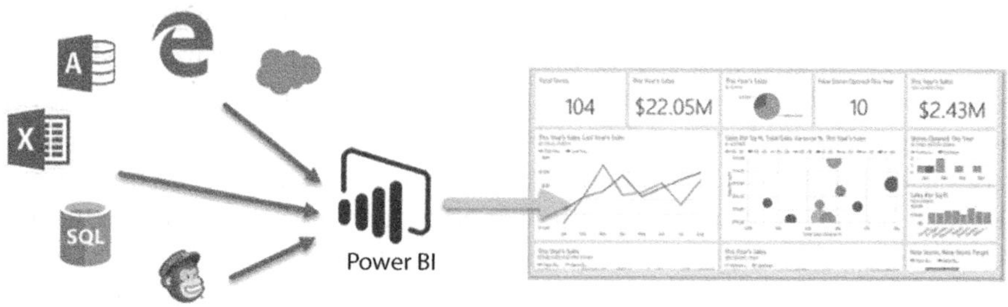

Figure 2.1 PowerBI collaborative environment.

Power BI has some different components available that can certainly be used separately like PowerBI DesktopPowerBU Service, PowerBI Mobile Apps, etc. (Figure 2.2).

No doubt the wide usability of PowerBI is due to the additional features that it provides over the existing tools used for analytics. Some add-ons include facilities like data warehousing, data discovery, and undoubtedly good interactive dashboards. The interface provided by PowerBI is both desktop based and cloud powered. Added to it its scalability ranges across the whole organization.

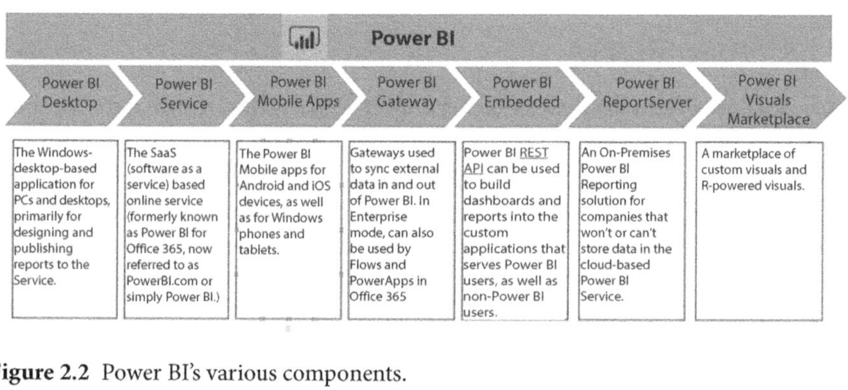

Figure 2.2 Power BI's various components.

Power BI is free, and initially, its analysis work begins with a desktop app where the reports are made then it is followed up by publishing them on Power BI service from where they can be shared over mobile where these reports can easily be viewed.

Power BI can either be used from the Microsoft store or downloading the software locally for the device. The Microsoft store version is an online form of this tool. Basic views like report view, data view, relationship view play a significant role in visualizations.

2.3.2 Soft Skills

It can be a tedious task to explain the technicalities behind the analysis part to a nontechnical audience. It is a crucial skill to be able to explain and communicate well what your data and related findings have to say or depict. As someone working on data you should have the ability to interpret data and thus impart the story it has to tell.

Along with technical skills, these soft skills play a crucial role. Just technical know-how cannot make you sail through swiftly, lest you possess the right soft skills to express that you cannot do justice to it.

As someone working with and on data you need to comfort the audience with your results and inform them how these results can be used and thereafter improve the business problem in a particular way. That is a whole lot of communicating. Here we will discuss a few of those skills that someone working in a Corporate must possess to ease things for themselves.

2.3.2.1 Presentation Skills

Presentation may look like an old way or tedious as well for that very matter but they are not going to go anywhere anytime soon. As a person working with data you are going to have to at some time or another to deliver a presentation. There are different approaches and techniques to effectively handle different classes of presentations:

One-on-One: A very intimate form of presentations whereby the delivery of information is to one person, i.e., a single stakeholder. Here the specific message is conveyed directly. It is important to make an effective engagement with the person whom the presentation is being given. The speaker should not only be a good orator but should also possess the ability to make an effective and convincing story which is supported by facts and figures so that it increases credibility.

Small Intimate Groups: This presentation is usually given to the board of members. These types of presentations are supposed to be short, sharp and to the point, because the board often has a number of topics on agenda. All facts and figures have to be precise and correct and that the number has been double checked. Here the meetings are supposed to end with a defined and clear conclusion to your presentation.

Classroom: It is a kind of presentation whereby you involve around 20 to 40 participants in your presentations, it becomes more complex to get involved with each and every attendee, hence make sure that whatever you say is precise and captivating. Here, it is the duty of the presenter to keep the message in his presentation very precise and relevant to what you have said. Make sure that your message is framed appropriately and when you summarize just inform them clearly with what you have presented.

Large Audiences: These types of presentation are often given at the conferences, large seminars and other public events. In most of the cases the presenter has to do brand building alongside conveying the message that you want to convey or deliver. It is also important to be properly presentable in terms of dressing sense. Use the 10-20-30 rule: 10 slides, 20pt font and 30 minutes. Make sure you are not just reading out the PPT. You will have to explain the presentations precisely to clarify the motive of your presentation. Do not try and squeeze in more than three to five key points. During a presentation, it should be you as a person who should be in the focus rather than the slides that you are presenting. And never, ever read off the slides or off a cheat sheet.

2.3.2.2 Storytelling

Storytelling is as important as giving presentations. Via storytelling the presenter basically makes their data speak and that is the most crucial task as someone working on data. To convey the right message behind your complex data, be it in terms of code or tool that you have used, the act of effective storytelling makes it simplified.

2.3.2.3 Business Insights

As an analyst, it is important that you have a business acumen too. You should be able to draw interpretations in context to business so that you facilitate the company's growth. Towards the end it is the aim of every company to use these insights to work on their market strategies so as to increase their profits. If you already possess them it becomes even easy to work with data and eventually be an asset to the organization.

2.3.2.4 Writing/Publishing Skills

It is important that the presenter must possess good writing and publishing skills. These skills are used for many purposes in a corporate world as an analyst. You might have to draft reports or publish white papers on your work and document them. You will have to draft work proposals or formal business cases for c-suite. You will be responsible to send official emails to the management. A corporate work culture does not really accept or appreciates social media slang. They are supposed to be well documented and highly professional. You might be responsible for publishing content on web pages.

2.3.2.5 Listening

Communication is not just about what you speak. It comprises both your speaking and listening skills. It is equally important to listen to what is the problem statement or issue that you are supposed to work on, so as to deliver the efficient solution. It is important to listen to what they have to say—what are their priorities, their challenges, their problems, and their opportunities. Make sure that everything you have to do should be able to deliver and communicate aptly. For this you first yourself have to understand them and analyze what can be the effect of different things on the business. As someone on data it is important that you should make constant efforts to perceive what is being communicated to you. As an effective listener you hear that is being said, assimilate and then respond accordingly. As an active listener you can respond by speaking what has been spoken so that you can cross check or confirm that you heard it right. As a presenter, you should show active interest in what others have to say. As an analyst you should be able to find important lessons from small things. They can act as a source of learning for you. Look for larger messages behind the data.

Data analysts should always be on the lookout for tiny mistakes that can lead to larger problems in the system and to later them beforehand so as to avoid bigger mishappenings in near future.

2.3.2.6 Stop and Think

This goes hand-in-hand with listening. The presenter is supposed to not be immediate with the response that he/she gives to any sort of verbal or written communications. You should never respond in a haste manner because once you have said something on the company's behalf on record you cannot take your words back. This should be specially taken into account on the soft cases or issues that might drive a negative reaction or feedback. It is absolutely fine and acceptable to think about an issue and respond to it thereafter. Taking time to respond is acceptable rather than giving a response without thinking.

2.3.2.7 Soft Issues

Not just your technical skills will help you make a sail through. It is important to acquaint oneself to the corporate culture and hence you should not only know how to speak but how much to speak and what all to speak. An individual should be aware of corporate ethics and then can in all help the whole organization to grow and excel. There are a number of soft issues that are to be taken into account while being at a workplace. Some of them are as follows:

- Addressing your seniors at the workplace with ethics and politely.
- One should try not to get involved in gossip in the office.
- One should always dress appropriately, i.e., much expected formals, specifically when there are important meetings with clients or higher officials from the office.
- One should always treat fellow team members with respect.
- You should possess good manners and etiquette.
- One should always make sure that they respect the audience's opinion and listen to them carefully.
- You should always communicate openly and with much honesty.
- You should been keen to learn new skills and things.

2.4 Responsibilities as Database Administrator

2.4.1 Software Installation and Maintenance

As a DBA, it is his/her duty to make the initial installations in the system and configure new Oracle, SQL Server etc databases. The system administrator also takes the onus of deployment and setting up hardware for the database servers and then the DBA installs the database software and configures it for use. The new updates and patches are also configured by a DBA for use. DBA also handles ongoing maintenance and transfers data to the new platforms if needed.

2.4.2 Data Extraction, Transformation, and Loading

It is the duty of a DBA to extract, transform and load large amounts of data efficiently. This large data has been extracted from multiple systems and is imported into a data warehouse environment. This external data is thereafter cleaned and is transformed so that it is able to fit in the desired format and then it is imported into a central repository.

2.4.3 Data Handling

With an increasing amount of data being generated, it gets difficult to monitor so much data and manage them. The databases which are in image/document/sound/audio-video format can cause an issue being unstructured data. Efficiency of the data shall be maintained by monitoring it and at same time tuning it.

2.4.4 Data Security

Data security is one of the most important tasks that a DBA is supposed to do. A DBA should be well aware of the potential loopholes of the database software and the company's overall system and work to minimize risks. After everything is computerized and depends on the system so it cannot be assured of hundred percent free from the attacks but opting the best techniques can still minimize the risks. In case of security breaches a DBA has authority to consult audit logs to see the one who has manipulated the data.

2.4.5 Data Authentication

As a DBA, it is their duty to keep a check of all those people who have access to the database. The DBA is one who can set the permissions and what type of access is given to whom. For instance, a user may have permission to see only certain pieces of information, or they may be denied the ability to make changes to the system.

2.4.6 Data Backup and Recovery

It is important for a DBA to be farsighted and hence keeping in mind the worst situations like data loss. For this particular task they must have a backup or recovery plan handy. Thereafter they must take the necessary actions and undergo needed practices to recover the data lost. There might be other people responsible for keeping a backup of the data but a DBA must ensure that the execution is done properly at the right time. It is an important task of a DBA to keep a backup of data. This will help them restore the data in case of any sort of sudden data loss. Different scenarios and situations require different types of recovery strategies. DBA should always be prepared for any kind of adverse situations. To keep data secure a DBA must have a backup over cloud or MS azure for SQL servers.

2.4.7 Security and Performance Monitoring

A DBA is supposed to have the proper insights of what is the weakness of the company's database software and company's overall system. This will certainly help them to minimize the risk for any issues that may arise in the near future. No system is fully immune to any sort of attacks, but if the best measures are implemented then this can be reduced to a huge extent. If an adverse situation of attack arises then in that case a DBA ought to consult audit logs to validate who has worked with the data in the past.

2.4.8 Effective Use of Human Resource

An effective administrator can be one who knows how to manage his human resource well. As a leader it is his/her duty to not only assign the tasks as per his members skill set and help them grow and enhance their skills. There are chances of internal mismanagement due to which, at times, it is the company or the output of the team indirectly which has to suffer.

2.4.9 Capacity Planning

An intelligent DBA is one who plans for things way before and keeps all situations in mind, so is the situation of capacity planning. A DBA must know the size of the database currently and what is the growth of the database in order to make future predictions about the needs of the future. Storage basically means the amount of space the database needs in server and backup space as well. Capacity refers to usage level. If a company is growing and keeps adding many new users, the DBA will be supposed to handle the extra workload.

2.4.10 Troubleshooting

There can be sudden issues that may come up with the data. For the issues that come up this way, DBA are the right people to be consulted at the moment. These issues can be quickly restoring the lost data or handling the issue with cre in order to minimize the damage, a DBA needs to quickly understand and respond to the problems when they occur.

2.4.11 Database Tuning

Monitoring the performance is a great way to get to know where the database is to be tweaked so as to operate efficiently. The physical configuration of the database, indexing of the database and the way queries are being handled all can have a dramatic effect on the database's performance. If we monitor it in a proper way, then the tuning of the system can be done just based on the application, not that we will have to wait for the issue to arise.

2.5 Concerns for a DBA [12]

- A responsible DBA also has to look into issues like security breach or attack. A lot of businesses in the UK have reported an attempt of data breach at least once in the last year. The bigger companies hold a lot of data and as a result the risk that the company might face from the cyber criminals is also very large. The possibility increases to 66% for medium-sized firms and 68% for large firms.
- A Company's database administrator could also put their employees' data at risk. A DBA is often warned over and over again that a company's employees' behavior can have a big impact on data security in their organization. The level of security of data can bind them with the organization for a longer time. It should be kept in mind that data security is a two-way street. Sensitive information about people in your company is just as valuable as your customers' data, therefore security procedures and processes have to be of top most priority for both employees' and customers' information.
- A DBA might have to look at things like DDoS attacks that a company might have to face. The type of attacks are the ones in which the attackers attack the machines or take down the whole network resources. The type of attacks can be temporary or may disrupt the internet. There is a fear that they might even lead to severe financial losses. These attacks may even lead to huge loss of finances to the company. In ample of these attacks the attacker has particularly dived into the person's wallet. A prediction says that by 2021, these attacks will cost the world over $5 Billion.
- A DBA needs to make sure that the company is abiding by the rules and regulations set by the government. At times companies try to surpass some important checks in order to maximize profits and thus put data security at stake. As different countries have different policies the organizations

are supposed to change their terms accordingly. This is the duty of DBA to make sure they abide by all the regulations.

In 2016, UK businesses were fined £3.2 million in total for breaching data protection laws.

- A DBA could be responsible for putting some confidential property or data that is supposed to be secretive at risk. Cybercrimes are not just restricted to financial losses but they also put intellectual property at risk.

In the UK, 20% of businesses admit they have experienced a breach resulting in material loss.

- In case your company's database is hit with a virus, DBA will have to cater to such sudden mishappenings. WannaCry, StormWorm and MyDoom are some softwares that have topped the list of being mass destructors.

According to research conducted by the UK Government's National Cyber Security Program, it has been found that 33% of all data breaches are a consequence of malicious software.

- It is important that the passwords that you keep for your account are not the same or easily identifiable. These passwords might be easy for us to memorize but it is risky at the same time because they can easily be cracked. Short passwords are highly vulnerable to be encoded by the attackers. Keep passwords in a way that they are mixture of both lower and upper case alphabets and have special symbols too.
- A company could also suffer damaging downtime. Often companies spend a lot on PR teams to maintain a good image in the corporate world. This is primarily done to keep a hold of good customers and at the same time eliminate competition. However just a single flaw or attack can put things upside down. This can damage the company's had earned reputation and this damage shall be irreplaceable.

It has been found that per minute loss can amount to as high as £6,000 due to an unplanned outage.

- A data breach act could hurt a company's reputation. It is very important for a company to maintain a positive image in the corporate world. Any damages to their image can significantly damage their business and future prospects.

According to 90% of CEOs, striving to rebuild commercial trust among stakeholders after a breach is one of the most difficult tasks to achieve for any company—regardless of their revenue.

- It might even happen that this might result in physical data loss. The physical data loss is something irreplaceable and amounts to huge losses.

2.6 Data Mishandling and Its Consequences

The mishandling of data is basically termed as data breaching. Data breach [13] refers to the process of stealing of the information. The information is taken from the systems by the attackers without any knowledge of the owner or company. The attack is carried out in an unauthorized and unlawful way. Irrespective of the company size, their data can be attacked by the attackers. The data attacked might be highly confidential and sensitive. Now being accessed by the wrong people might lead to serious trade threats or security threats. The effects of the data breach can not only be harmful to the people whose data is at risk but can significantly damage the reputation of the company as well. Victims might even suffer serious financial losses in case they are related to credit card or passwords. A recent survey found out that the personal information stolen was at the first position followed up by financial data being stolen. This was evaluated on the data from year 2005 to 2015.

Data leaks are primarily malware attacks but at the same time there can be other factors too at the same time:

- Some insiders from the organization might have leaked the data.
- Fraud activities associated with Payments cards.
- Data loss could be another reason. This is primarily caused by mishandling.
- Unintended disclosure.

Data theft is continuing [14] to make headlines despite the fact that there has been a lot of awareness among the people and companies. Not just the awareness but there have been stricter laws formulated by the government to prevent the data breach activities. However, cybercriminals have found their way into people's data and have been posing a threat continuously. They have their different ways of getting into the network. It can be either through social engineering. technique or maybe malware or supply chain attacks. The attackers basically try to infiltrate profits via this.

Unfortunately, the main concern here is despite the repeated increase in issues of data breaching and threat to the data some organizations are certainly not prepared to handle these situations of attack on their systems. Many organizations are willingly underprepared and fail to inculcate proper security systems in their working to avert any sort of cyberattacks.

In a recent survey conducted, it was discovered that nearly 57% companies still do not have cyber security policy and this has risen to nearly 71% of the medium sized businesses having nearly 550 to 600 employees. Companies need to ponder on the after effects of data breach on them and their customers this will certainly compel them to work on bettering their system to avert the cyberattacks.

2.6.1 Phases of Data Breaching

- Research: This is the first and the foremost thing that an attacker would do. After having picked the target an attacker would find the necessary details needed for carrying out the activity/act of data breaching. They find the loopholes in the system or weakness which makes it easy for them to dive in the required information. They get the detail information about the company's infrastructure and do the primary stalking about the employees from various platforms.
- Attack: After having the much-needed details of the company and its infrastructure, the attacker makes/carries out the first step by making some initial contact either via some network or maybe social media.

In a network-based attack, the main task/purpose of the attacker is to exploit the weaknesses of the target's infrastructure to undergo the breach. The attackers may undergo an SQL injection or maybe session hijacking.

In a social attack, the attacker uses social engineering tactics to dive into the target network. They may hit the company's employees in the form of a well-crafted email, and thereafter, the email can phish data by compelling them to provide the personal data in the email. Or that mail may contain some malware attached to it that may get executed as and when the mail is opened.

Exfiltrate: So as soon as the attacker accesses the network they are free to extract any sort of information from the company's database. That data can be used by the attackers for any of the unlawful practices that will certainly harm the company's reputation and keep future prospects at stake.

2.6.2 Data Breach Laws

It is important to have administration intervention to prevent the ill practices that occur with data. Data breach laws and the related punishments vary in various nations. Many countries still do not require organizations to notify authorities in cases of data breach. In countries like the US, Canada, and France, organizations are obliged to notify affected individuals under certain conditions.

2.6.3 Best Practices For Enterprises

- Patch systems and networks accordingly. It is the duty of the IT administrators to make sure that the systems in the network are well updated. This will prevent them from the attackers and make them less vulnerable to being attacked in near future.
- Educate and enforce. It is crucial to keep the employees informed about the threats and at the same time impart the right knowledge to them regarding social engineering tactics. This way they will get acquainted with situations where they might have to handle any adverse situation.
- Implement security measures. Experimenting and implementing changes is the primary job here. They are supposed to identify risk factors, ponder on its solutions and then thereafter implement the measures. They also have to bring improvisations to keep a check of the solutions they have implemented.
- Create contingencies. It is crucial to be prepared for the worst so for this particular thing there should be an effective recovery plan put forward so that whenever there is a data breach activity the team and the people know do they handle it, who all will be the contact persons, what are the disclosure strategies, what all would be the mitigation steps and also that employees are well aware of this plan.

2.7 The Long-Term Consequences: Loss of Trust and Diminished Reputation

The long-term effect of data breach could be the loss of faith amongst the customers. The customers share their sensitive information with the company considering the fact that the company will certainly look into data security and their information with the company is safe. In a survey conducted in 2017 by PwC it was found that nearly 92% of people agreed with the fact that the companies will take customer's data security as a prime concern and top most priority. A goodwill of the company among the customers is highly valued and is the most prized asset. However, instances of data breach can be significantly harmful and damage the reputation earned with much effort and years of service with excellence.

The PwC [15] report found that 85% of consumers will not shop at a business if they have concerns about their security practices. In a study done by Verizon in year 2019, it was found that nearly 29% of people will not return to the company back again where they have suffered any sort of data breach. It is important to understand the consequences because this way the companies will be able to secure their businesses in the long run and at the same time maintain the reputation as well.

2.8 Solution to the Problem

Acknowledging critical data is the first step: As an administrator you cannot secure something that you do not acknowledge. Just have a look at your data about where it is located or how it is being stored or handled. You must look at it from an outsider's perspective. You must look at it from obvious places that are being overlooked such as workstations, network stations and backups. But at the same time there can be other areas too where data might be stored out of our security control such as cloud environments. All it takes is one small oversight to lead to big security challenges.

2.9 Case Studies

2.9.1 UBER Case Study [16]

Before we get into knowing about how UBER used data analytics in improving and optimizing their business, let us make an effort to understand the business model of UBER and understand its work.

Uber is basically a digital aggregator application platform, which connects the passengers who need to commute from one place to another with drivers who are interested in providing them the pick and drop facility. The demand/need is put forward by the drivers and drivers supply the demand. Also Uber at the same time acts as a facilitator to bridge the gap and make this hassle free process via a mobile application. Let us study key components of UBER's working model by understanding the following chart:

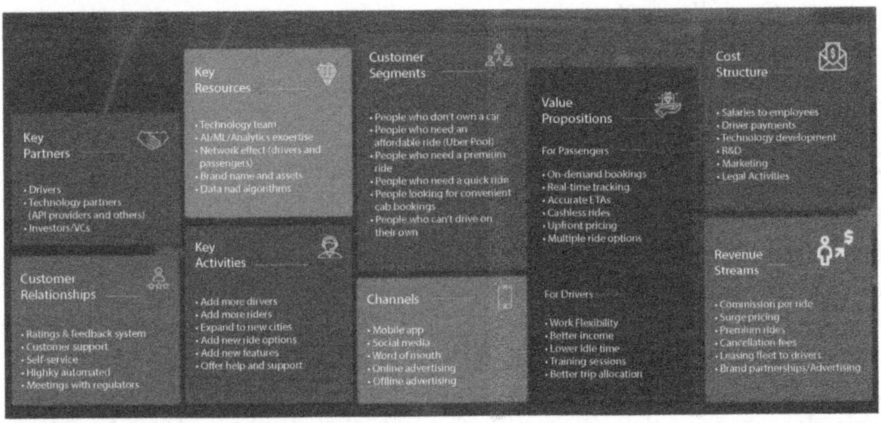

Figure 2.3 UBER's working model.

Since riders and drivers are the crucial and most important part of UBER's business model (Figure 2.3). UBER certainly has valuable features to provide its users/riders, some of which are:

- Booking cab on demand
- Tracking the movement in real time
- Precise Estimated Time of Arrival
- Using digital media for paying money cashless way
- Lessen the waiting time
- Upfront ride fares
- Ample of Cab Options

Similarly, Uber's value propositions for drivers are:

- Flexibility in driving on their own conditions and terms
- Better Compensation in terms of money earned
- Lesser idle time to get new rides
- Better trip allocation

The main issue that pops up now is that how is Uber deriving its monetary profits? Or what is the system by which Uber streams its revenue? If we have a look at it from a higher view, Uber basically takes commission from drivers for every ride that is being booked from their app also at the same time they have different ways to increase revenue for themselves:

- Commission from rides
- Premium rides
- Surge pricing
- Cancellation fees
- Leasing cars to drivers
- Uber eats and Uber freights

2.9.1.1 Role of Analytics and Business Intelligence in Optimization

Uber undoubtedly has a huge database of drivers, so whenever a request is put in for a car, the Algorithm is put to work and it will associate you to the nearest drive in your locality or area. In the backend, the company's system stores the data for each and every journey taken—even if there is no passenger in the car. The data is henceforth used by business teams to closely study to draw interpretations of supply and demand market forces. This also supports them in setting fares for the travel in a given location. The team of the company also studies the way transportation systems are being managed in different cities to adjust for bottlenecks and many other factors that may be an influence.

Uber also keeps a note of the data of its drivers. The very basic and mere information is not just what Uber collects about its drivers but also it also monitors their speed, acceleration, and also monitors if they are not involved with any of their competitors as well for providing any sort of services.

All this information is collected, crunched and analyzed to carry forward some predictions and devise visualizations in some vital domains namely customer wait time, helping drivers to relocate themselves in order to take advantage of best fares and find passengers accordingly at the right rush hour. All these items are implemented in real time for both drivers and passengers alike.

The main use of Uber's data is in the form of a model named "Gosurge" for surge pricing." Uber undergoes real-time predictive modeling on the basis of traffic patterns, supply and demand.

If we look at it from a short term point of view, surge pricing substantially has a vital effect in terms of the rate of demand, while long-term use could be using the service for customer retention or losing them. It has effectively made use of Machine Learning for the purpose of price prediction especially in case of price hiking, thus it can effectively increase the adequate price to meet that demand, and surge can also be reduced accordingly. This is primarily because the customer backlash is strong in case of rate hiking.

Keeping in mind, these parameters of supply and demand keep varying from city to city, so Uber engineers have found a way to figure out the "pulse" of a city to connect drivers and riders in an efficient manner. Also we have to keep this factor in mind that not all metropolitan cities are alike. Let us see an overview comparison of London and New York for a better overview and insights:

Collection of all the information is basically one minor step in the long journey of Big data and drawing further interpretations from the same. The real question is - How can Uber channelize this huge amount of data to make decisions and use this information? How do they basically glean main points to ponder on out of such a huge amount of data? For example, How does Uber basically manage millions of GPS locations. Every minute, the database is getting filled with not just driver's information but also it has a lot of information related to users. How does Uber make effective use of the very minute details so as to better manage the moving fellas and things from one place to another ?

Their answer is data visualization.

Data visualization specialists have a varied variety of professionals from Computer Graphics background to information design (Figure 2.4). They look into different aspects right from Mapping and framework developments to data that the public sees. And a lot of these data explorations and visualizations are completely fresh and never have been done before. This has basically developed a need for tools to be developed in-house.

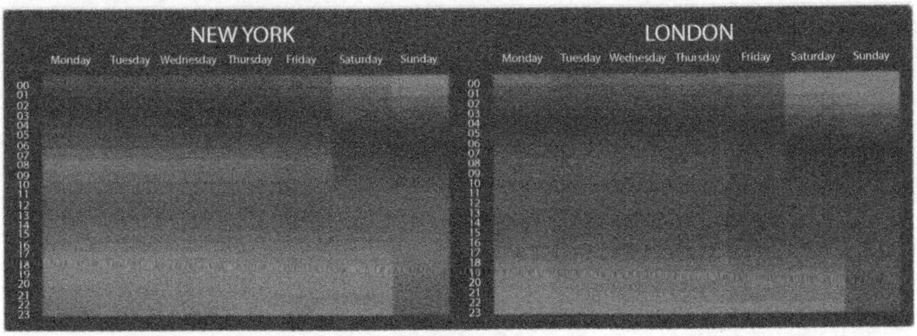

When Uber trips occur throughout the week in New York City and London. The bridge brightness levels per hour and day are compared to the city itself. All times are standardized to the local time zone and expressed in military time (i.e. 20 is 20:00, or 8 pm).

Figure 2.4 UBER's trip description in a week.

2.9.1.2 Mapping Applications for City Ops Teams

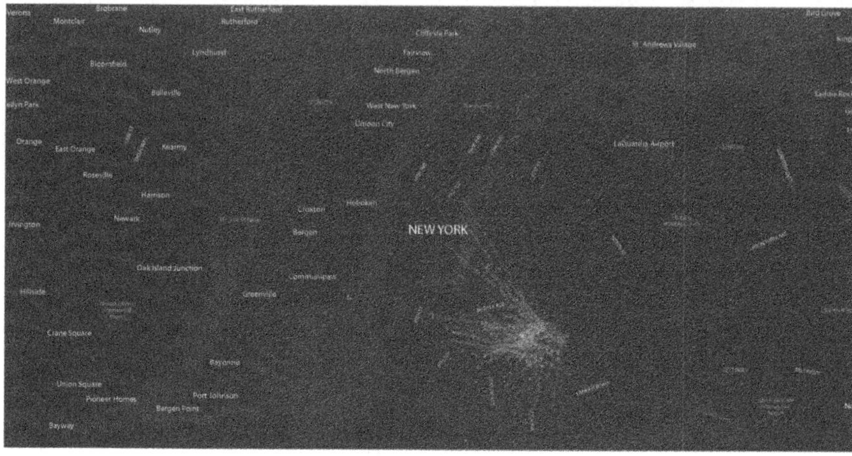

Figure 2.5 UBER's city operations map visualizations.

These visualizations are not the ones specifically meant for the data scientist or engineers but in general for the public as well for having a better understanding and clarity (Figure 2.5). It helps the public to better understand the working insights of the giant, such as visualization helps to know about uberPOOL and thus plays a significant role to reduce traffic (Figure 2.6).

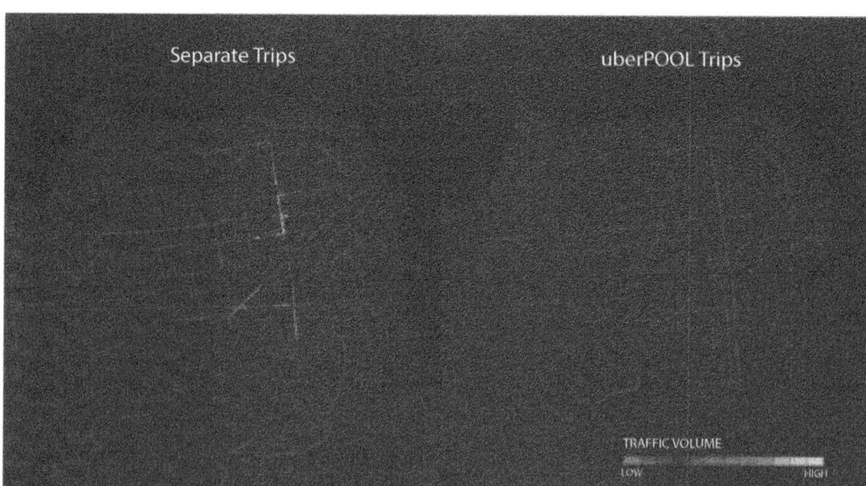

Figure 2.6 UBER's separate trips and UBER-Pool trips.

Another example of this visualization is particularly in Mega cities, where understanding the population density of a given area is of significant importance and they play a vital role in dynamic pricing change. Uber illustrates this with a combination of layers that helps them to narrow down to see a specific are in depth (Figure 2.7):

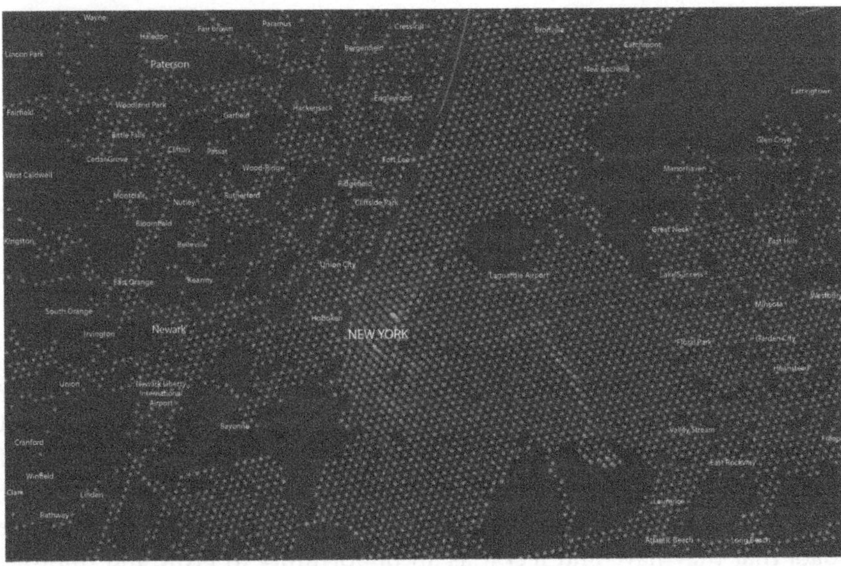

Figure 2.7 Analysis area wise in New York.

Not just visualizations, Forecasting as well plays a significant role in business intelligence techniques that are being used by Uber to optimize future processes.

2.9.1.3 Marketplace Forecasting

A crucial element of the platform, marketplace forecasting helps Uber to predict user supply and demand in a spatiotemporal fashion to help the drivers to reach the high-demand areas before they arise, thereby increasing their trip count and hence shooting their earnings (Figure 2.8). Spatiotemporal forecasts are still an open research area.

Figure 2.8 Analysis area wise in spatiotemporal format.

2.9.1.4 Learnings from Data

It is just one aspect to describe how Uber uses data science, but another aspect is to completely discover what these results or findings have to say beyond that particular thing. Uber teaches us an important thing to not just have a store of humongous data but at the same time making a use of it effectively. Also an important takeaway from the working style of Uber is that they have a passion to drive some useful insights from every ounce of data that they have and feel it as an opportunity to grow and improve the business.

It is also worth realizing that it is crucial to explore and gather data independently and analyze it for what it is and what is actually going to make insights come up.

2.9.2 PepsiCo Case Study [17]

PepsiCo primarily depends on the huge amount of data to supply its retailers in more than 200 countries and serve a billion customers every day.

Supply cannot be made over the designated amount because it might lead to the wastage of resources. Supplying a little amount as well is problematic because it shall affect the profits and loss and company may reconcile with unhappy and dissatisfied retailers. An empty shelf also paves a way for customers to choose the competitor's product, which is certainly not a good sign added to it, it has long-term drawbacks on the brand.

Now PepsiCo mainly uses data visualizations and analysis to forecast the sales and make other major decisions. Mike Riegling works as an analyst with PepsiCo in the CPFR team. His team provides insights to the sales and management team. They collaborate with large retailers to provide the supply of their products in the right quantity for their warehouses and stores.

"The journey to analytics success was not easy. There were many hurdles along the way. But by using Trifacta to wrangle disparate data" says Mike. Mike and his teammates made significant reductions to reduce the end-to-end run time of the analysis by nearly 70% Also adding Tableau to their software usage it could cut report production time as much as 90%.

"It used to take an analyst 90 minutes to create a report on any given day. Now it takes less than 20 minutes," says Mike.

2.9.2.1 Searching for a Single Source of Truth

PepsiCo's customers give insights that consist of warehousing inventory, store inventory and point-of-sale inventory. The company then rechecks this data with their own shipping history, produced quantity, and further forecast data.

Every customer has their own data standards. It was difficult for the company to do data wrangling in this case. It could take a long time, even months at times, to generate reports. It was another important task for them to derive some significant sales insights from these reports and data. Their teams initially used only Excel to analyze data of large quantities which is primarily messy. At the same time, the team had no proper method to spot errors. A missing product at times led to huge errors in reports and in accurate forecasts. This could lead to losses as well.

2.9.2.2 Finding the Right Solution for Better Data

The most important task for the company initially was to bring coherence to their data. For this they used Tableau and thereafter results were in the form of improved efficiency. Now the new reports basically run without much involvement of multiple access and PepsiCo servers and they run directly on hadoop. The analysts could make manipulations using trifacta now.

As per what company's officials have said the has been successfully able to bridge the gap between Business and Technology. This technology has successfully helped them to access the raw data and do the business effectively. The use of technology has been such a perfect blend that it has been able to provide a viable solution to each of their problems in an effective way Tableau provides them with finishing step, i.e., basically providing with powerful analytics and interactive visualizations, helping the businesses to draw insights from the volumes of data. Also the analysts as PepsiCo share their reports on business problems with the management using Tableau Servers.

2.9.2.3 Enabling Powerful Results with Self-Service Analytics

Now in the case of PepsiCo it was the combined use of various tools namely Tableau, Hortonworks and Trifacta that have played a vital role driving the key decisions taken by analytic teams. They have helped CPFR teams drive the business forward and thus increased the customer orders. The changes were also visible clearly.

This process of using multiple analytics tools has had multifaceted advantages. Not just it has reduced the time invested upon the data for preparation but added to it ;it has increased an overall data quality.

The use of technology has been of great use for the company. This has been able to save their time significantly as they have been investing time analyzing the data and making their data tell a relevant story rather than putting their data together. They have been able to form better graphs now and study them effectively with much accuracy.

PepsiCo has successfully been able to turn customer data around and successfully present it to the rest of the company such that everyone can understand it better than their competitors.

2.10 Conclusion

This chapter concludes by making the readers aware of both technical and nontechnical skills that they must possess to work with data. The skills will help readers to be effective in dealing with data and grow professionally. Also it makes them aware of their responsibilities as a data or database administrator. Toward the end, we throw some light upon the consequences of data mishandling and how to handle such situations.

References

1. https://www.geeksforgeeks.org/difference-between-data-administrator-da-and-database-administrator-dba/ [Date: 11/11/2021]
2. https://searchenterpriseai.techtarget.com/definition/data-scientist [Date: 11/11/2021]

3. https://whatisdbms.com/role-duties-and-responsibilities-of-database-administrator-dba/ [Date: 11/11/2021]
4. https://www.jigsawacademy.com/blogs/data-science/dba-in-dbms/ [Date: 11/11/2021]
5. https://www.jigsawacademy.com/blogs/data-science/dba-in-dbms/ [Date: 11/11/2021]
6. http://www.aaronyeo.org/books/Data_Science/Python/Wes%20McKinney%20-%20Python%20for%20Data%20Analysis.%20Data%20Wrangling%20with%20Pandas,%20NumPy,%20and%20IPython-O%E2%80%99Reilly%20(2017).pdf [Date: 11/11/2021]
7. https://www3.nd.edu/~kkelley/publications/chapters/Kelley_Lai_Wu_Using_R_2008.pdf [Date: 11/11/2021]
8. https://reader.elsevier.com/reader/sd/pii/S2212567115000714?token=7721440CD5FF27DC8E47E2707706E08A6EB9F0FC36BDCECF1D3C687635F5F1A69B809617F0EDFFD3E3883CA541F0BC35&originRegion=eu-west-1&originCreation=20210913165257 [Date: 11/11/2021]
9. https://towardsdatascience.com/introduction-to-scala-921fd65cd5bf [Date: 11/11/2021]
10. https://www.softwebsolutions.com/resources/tableau-data-visualization-consulting.html [Date: 11/11/2021]
11. https://www.datacamp.com/community/tutorials/data-visualisation-powerbi [Date: 11/11/2021]
12. https://dataconomy.com/2018/03/12-scenarios-of-data-breaches/ [Date: 11/11/2021]
13. https://www.trendmicro.com/vinfo/us/security/definition/data-breach [Date: 11/11/2021]
14. https://www.metacompliance.com/blog/5-damaging-consequences-of-a-data-breach/ [Date: 11/11/2021]
15. https://www.pwc.com/us/en/advisory-services/publications/consumer-intelligence-series/protect-me/cis-protect-me-findings.pdf [Date: 11/11/2021]
16. https://www.skillsire.com/read-blog/147_data-analytics-case-study-on-optimizing-bookings-for-uber.html [Date: 11/11/2021]
17. https://www.tableau.com/about/blog/2016/9/how-pepsico-tamed-big-data-and-cut-analysis-time-70-59205 [Date: 11/11/2021]

3

Data Wrangling Dynamics

Simarjit Kaur*, Anju Bala and Anupam Garg

Department of Computer Science and Engineering, Thapar Institute of Engineering and Technology, Patiala, India

Abstract

Data is one of the prerequisites for bringing transformation and novelty in the field of research and industry, but the data available is unstructured and diverse. With the advancement in technology, digital data availability is increasing enormously and the development of efficient tools and techniques becomes necessary to fetch meaningful patterns and abnormalities. Data analysts perform exhaustive and laborious tasks to make the data appropriate for the analysis and concrete decision making. With data wrangling techniques, high-quality data is extracted through cleaning, transforming, and merging data. Data wrangling is a fundamental task that is performed at the initial stage of data preparation, and it works on the content, structure, and quality of data. It combines automation with interactive visualizations to assist in data cleaning. It is the only way to construct useful data to further make intuitive decisions. This paper provides an overview of data wrangling and addresses challenges faced in performing the data wrangling. This paper also focused on the architecture and appropriate techniques available for data wrangling. As data wrangling is one of the major and initial phases in any of the processes, leading to its usability in different applications, which are also explored in this paper.

Keywords: Data acquisition, data wrangling, data cleaning, data transformation

3.1 Introduction

Organizations and researchers are focused on exploring the data to unfold hidden patterns for analysis and decision making. A huge amount of data has been generated every day, which organizations and researchers

**Corresponding author*: skaur60_phd19@thapar.edu

M. Niranjanamurthy, Kavita Sheoran, Geetika Dhand, and Prabhjot Kaur (eds.) Data Wrangling: Concepts, Applications and Tools, (53–70) © 2023 Scrivener Publishing LLC

gather. Data gatheredor collected from different sources such as databases, sensors, surveys is heterogeneous in nature that contains multiple file formats. Initially, this data is raw and needs to be refined and transformed to make it applicable and serviceable. The data is said to be credible if it is recommended by data scientists and analysts and provides valuable insights [1]. Then thedata scientist's job starts, and several data refinement techniques and tools have been deployed to get meaningful data.

The process of data acquisition, merging, cleaning, and transformation is known as data wrangling [2]. The data wrangling process integrates, transforms, clean, and enrich the data and provides an enhanced quality dataset [3]. The main objective is to construct usable data, to convert it into a format that can be easily parsed and manipulated for further analysis. The usefulnessof data has been assessed based on the data processing tools such as spreadsheets, statistics packages, and visualization tools. Eventually, the output should be the original representation of the dataset [4]. Future research direction should focus on preserving the data quality and providing efficient techniques to make data usable and reproducible. The subsequent section discusses the research done by several researchers in the field of data wrangling.

3.2 Related Work

As per the literature reviewed, many researchers proposed and implemented data wrangling techniques. Some of the relevant works done by researchers have been discussed here. Furche *et al.* [5] proposed an automated data wrangling architecturebased on the concept of Extract, Transform and Load (ETL) techniques. Data wrangling research challenges and the need to propose techniques to clean and transform the data acquired from several sources. The researchers must provide cost-effective manipulations of big data. Kandel *et al.* [6] presented research challenges and practical problems faced by data analysts to create quality data. In this paper, several data visualization and transformation techniques have been discussed. The integration of visual interface and automated data wrangling algorithms provide better results.

Braun *et al.* [7] addressed the challenges organizational researchers face for the acquisition and wrangling of big data. Various sources of significant data acquisition have been discussed, and the authors have presented data wrangling operations applied for making data usable. In the future, data scientists must consider and identify how to acquire and wrangle big data efficiently. Bors *et al.* [8] proposed an approach for exploring data, and a visual analytics approach has been implemented to capture the data from data wrangling operations. It has been concluded that various data wrangling operations have a significant impact on data quality.

Barrejon *et al.* [9] has proposed a model based on sequential heterogeneous incomplete variational autoencoders for medical data wrangling operations. The experimentation has been performed on synthetic and real-time datasets to assess the model's performance, and the proposed model has been concluded as a robust solution to handle missing data. Etati *etal.* [10] deployed data wrangling operations using power BI query editor for predictive analysis. Power query editor is a toolthat has been used for the transformation of data. It can perform data cleaning, reshaping, and data modeling by writing R scripts. Data reshaping and normalization have been implemented.

Rattenbury *et al.* [11] have provided a framework containing different data wrangling operations to prepare the data forfurther and insightful analysis. It has covered all aspects of data preparation, starting from data acquisition, cleaning, transformation, and data optimization. Various tools have been available, but the main focus has been on three tools: SQL, Excel, and Trifacta Wrangler. Further, these data wrangling tools have been categorized based on the data size, infrastructure, and data structures supported. The tool selection has been made by analyzing the user's requirement and the analysis to be performed on the data. However, several researchers have done much work, but still, there are challenges in data wrangling. The following section addresses the data wrangling challenges.

3.3 Challenges: Data Wrangling

Data wrangling is a repetitious process that consumes a significant amount of time. The time-intensive nature of data wrangling isthe most challenging factor. Data scientists and analysts say that it takes almost 80% of the time of the whole analysis process [12]. The size of data is increasing rapidly with the growth of information and communication technology. Due to that, organizations have been hiring more technical employees and putting their maximum effort into data preparation, and the complex nature ofdata is a barrier to identify the hidden patterns present in data. Some of the challenges of data wrangling have been discussed as follows:

- The real time data acquisition is the primary challenge faced by data wrangling experts. The data entered manually may contain errors such as the unknown values at a particular instance of time can be entered wrongly. So the data collected should record accurate measurements that can be further utilized for analysis and decision making.

- Data collected from different sources is heterogeneous that contains different file formats, conventions, and data structures. The integration of such data is a critical task, so incompatible formats and inconsistencies must be fixed before performing data analysis.
- As the amount of data collected over time grows enormously, efficient data wrangling techniques could only process this big data. Also, it becomes difficult to visualize raw data to extract abnormalities and missing values.
- Many transformation tasks have been deployed on data, including extraction, splitting, integration, outlier elimination, and type conversion. The most challenging task is data reformatting and validating required by transformations. Hence data must be transformed into the attributes and features which can be utilized for analysis purposes.
- Some data sources have not provided direct access to data wranglers; due to that, most of the time has been wasted in applying instructions to fetch data.
- The data wrangling tools must be well understood to select appropriate tools from the available tools. Several factors such as data size, data structure, and type of infrastructure influence the data wrangling process. However, these challenges must be addressed and resolved to perform effective data wrangling operations. The subsequentsection discusses the architecture of data wrangling.

3.4 Data Wrangling Architecture

Data wrangling is called the most important and tedious step in data analysis, but data analysts have ignored it. It is the process of transforming the data into usable and widely used file formats. Every element of data has been checked carefully or eliminated if it includes inconsistent dates, outdated information, and other technological factors. Finally, the data wrangling process addresses and extracts the most fruitful information present in the data. Data wrangling architecture has been shown in Figure 3.1, and the associated steps have been elaborated as follows:

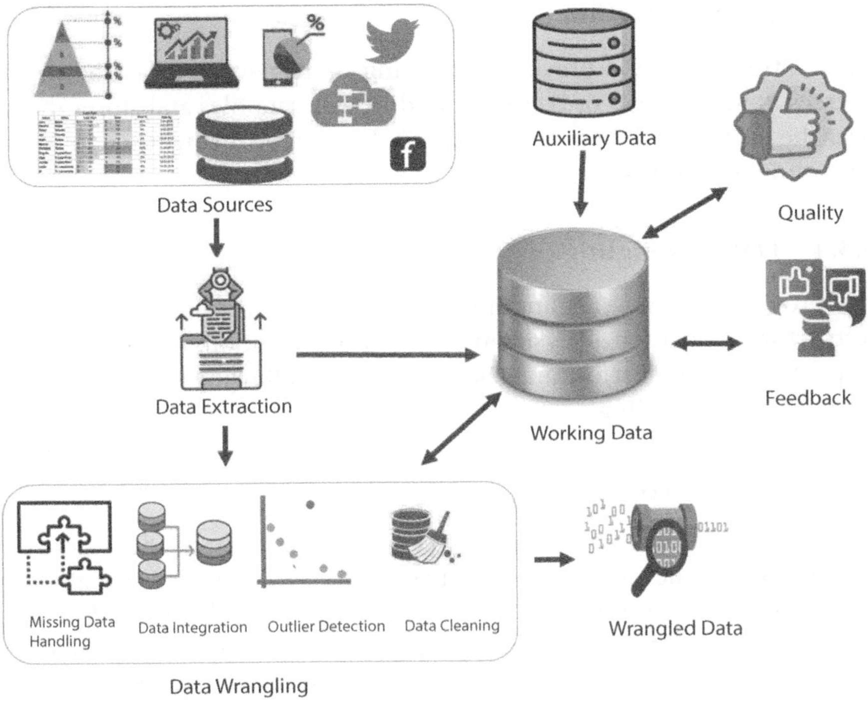

Figure 3.1 Graphical depiction of the data wrangling architecture.

3.4.1 Data Sources

The initial location where the data has originated or been produced is known as the data source. Data collected from different sources contain heterogeneous data having other characteristics. The data source can be stored on a disk or a remote server in the form of reports, customer or product reviews, surveys, sensors data, web data, or streaming data. These data sources can be of different formats such as CSV, JSON, spreadsheet, or database files that other applications can utilize.

3.4.2 Auxiliary Data

The auxiliary data is the supporting data stored on the disk drive or secondary storage. It includes descriptions of files, sensors, data processing, or the other data relevant to the application. The additional data required can be the reference data, master data, or other domain-related data.

3.4.3 Data Extraction

Data extraction is the process of fetching or retrieving data from data sources. It also merges or consolidates different data files and stores them near the data wrangling application. This data can be further used for data wrangling operations.

3.4.4 Data Wrangling

The process of data wrangling involves collecting, sorting, cleaning, and restructuring data for analysis purposes in organizations. The data must be prepared before performing analysis, and the following steps have been taken in data wrangling:

3.4.4.1 Data Accessing

The first step in data wrangling is accessing the data from the source or sources. Sometimes, data access is invoked by assigning access rights or permissions on the use of the dataset. It involves handling the different locations and relationships among datasets. The data wrangler understands the dataset, what the dataset contains, and the additional features.

3.4.4.2 Data Structuring

The data collected from different sources has no definite shape and structure, so it needs to be transformed to prepare it for the data analytic process. Primarily data structuring includes aggregating and summarizing the attribute values. It seems a simple process that changes the order of attributes for a particular record or row. But on the other side, the complex operations change the order or structure of individual records, and the record fields have been further split into smaller components. Some of the data structuring operations transform and delete few records.

3.4.4.3 Data Cleaning

Data cleaning is also a transformation operation that resolves the quality and consistency of the dataset. Data cleaning includes the manipulation of every field value within records. The most fundamental operation is handling the missing values. Eventually, raw data contain many errors that should be sorted out before processing and passing the data to the next stage. Data cleaning also involves eliminating the outliers, doing corrections, or deleting abnormal data entirely.

3.4.4.4 *Data Enriching*

At this step, data wranglers become familiar with the data. The raw data can be embellished and augmented with other data. Fundamentally, data enriching adds new values from multiple datasets. Various transformations such as joins and unions have been deployed to combine and blend the records from multiple datasets. Another enriching transformation is adding metadata to the dataset and calculating new attributes from the existing ones.

3.4.4.5 *Data Validation*

Data validation is the process to verify the quality and authenticity of data. The data must be consistent after applying data-wrangling operation. Different transformations have been applied iteratively and the quality and authenticity of the data have been checked.

3.4.4.6 *Data Publication*

On the completion of the data validation process, data is ready to be published. It is the final result of data wrangling operations performed successfully. The data becomes available for everyone to perform analysis further.

3.5 Data Wrangling Tools

Several tools and techniques are available for data wrangling and can be chosen according to the requirement of data. There is no single tool or algorithm that suits different datasets. The organizations hire various data wrangling experts based on the knowledge of several statistical or programming languages or understanding of a specific set of tools and techniques. This section presents popular tools deployed for data wrangling:

3.5.1 Excel

Excel is the 30-year-old structuring tool for data refinement and preparation. It is a manual tool used for data wrangling. Excel is a powerful and self-service tool that enhances business intelligence exploration by providing data discovery and access. The following Figure 3.2 shows the missing values filled by using the random fill method in excel. The same column data is used as a random value to replace one or more missing data values in the corresponding column. After preparing the data, it can be deployed

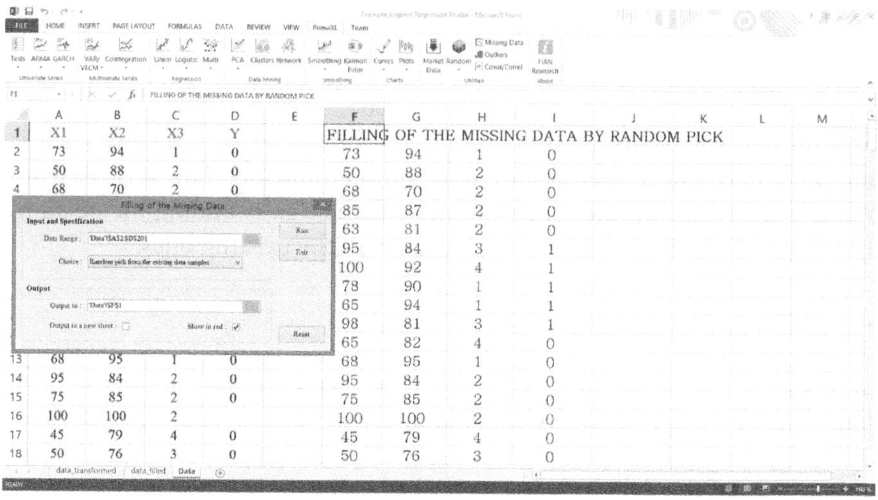

Figure 3.2 Image of the Excel tool filling the missing values using the random fill method.

for training and testing any machine learning model to extract meaningful insights out of the data values.

3.5.2 Altair Monarch

Altair Monarch is a desktop-based data wrangling tool having the capability to integrate the data from multiple sources [16]. Data cleaning and several transformation operations can be performed without coding, and this tool contains more than 80 prebuilt data preparation functions.

Altair provides graphical user interface and machine learning capabilities to recommend data enrichment and transformations. The above Figure 3.3 shows the initial steps to open a data file from different sources. First, click on Open Data to choose the data source and search the required file from the desktop or other locations in the memory or network. The data can also be download from the web page and drag it to the start page. Further, data wrangling operations can be performed on the selected data, and prepared data can be utilized for data analytics.

3.5.3 Anzo

Anzo is a graph-based approach offered by Cambridge Semantics for exploring and integrating data. Users can perform data cleaning, data blending operations by connecting internal and external data sources.

Figure 3.3 Image of the graphical user interface of Altair tool showing the initial screen to open a data file from different sources.

The user can add different data layers for data cleansing, transformation, semantic model alignment, relationship linking, and access control operation [19]. The data can be visualized for understanding and describing the data for organizations or to perform analysis. The features and advantages of Anzo Smart Data Lake have been depicted in the following Figure 3.4. It connects the data from different sources and performs data wrangling operations.

3.5.4 Tabula

Tabula is a tool for extracting the data tables out of PDF files as there is no way to copy and paste the data records from PDF files [17]. Researchers use it to convert PDF reports into Excel spreadsheets, CSVs, and JSON files, as shown in Figure 3.5, and further utilized in analysis and database applications.

3.5.5 Trifacta

Trifacta is a data wrangling tool that contains a suite of three iterations: Trifacta Wrangler, Wrangler Edge, and Wrangler Enterprise. It supports various data wrangling operations such as data cleaning, transformation without writing codes manually [14]. It makes data usable and accessible

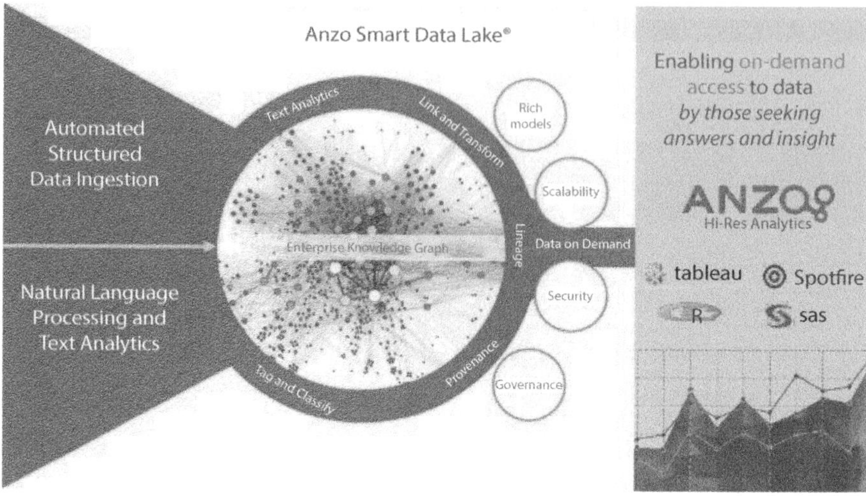

Figure 3.4 Pictorial representation of the features and advantages of Anzo Smart Data Lake tool.

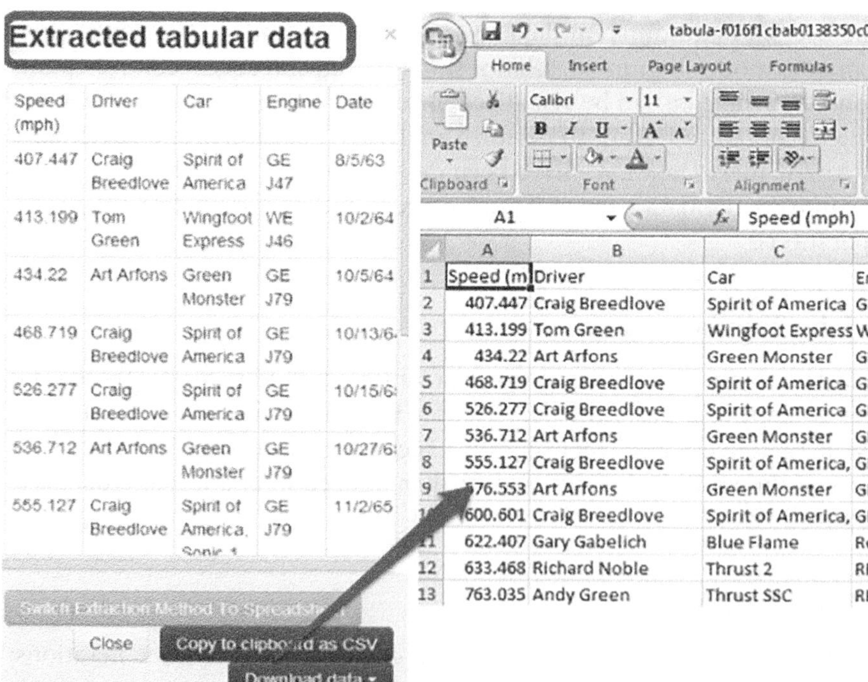

Figure 3.5 Image representing the interface to extract the data files in .pdf format to other formats, such as .xlsx, .csv.

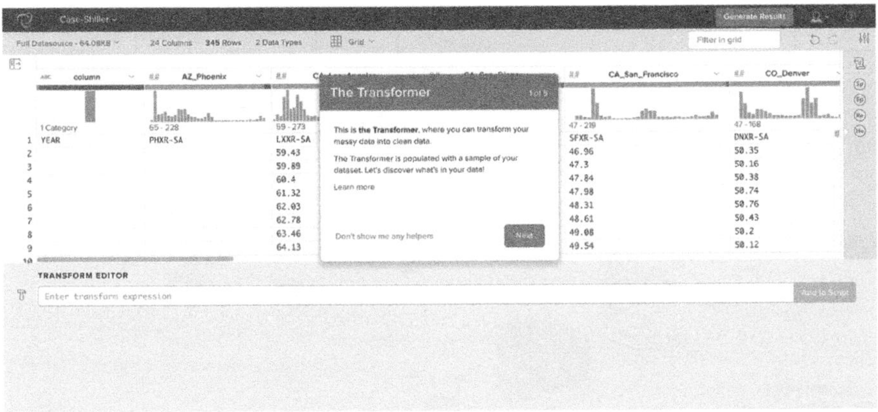

Figure 3.6 Image representing the transformation operation in Trifacta tool.

that suits to requirements of anyone. It can perform data structuring, transforming, enriching, and validation. The transformation operation is depicted in Figure 3.6. The users of Trifacta are facilitated with preparing and cleaning data; rather than mailing the excel sheets, the Trifacta platform provides collaboration and interactions among them.

3.5.6 Datameer

Datameer provides a data analytics and engineering platform that involves data preparation and wrangling tasks. It offers an intuitive and interactive spreadsheet-style interface that facilitates the user with functions like transform, merge and enrich the raw data to make it a readily used format [13]. Figure 3.7 represents how Datameer accepts input from heterogeneous data sources such as CSV, database files, excel files, and the data files from web services or apps. There is no need for coding for cleaning or transforming the data for analysis purposes.

3.5.7 Paxata

Paxata is a self-service data preparation tool that consists of an Adaptive Information Platform. It is a flexible product that can be deployed quickly and provides a visual user interface similar to spreadsheets [18]. Due to it, any user can utilize this tool without learning the tool entirely. Paxata is also enriched with Intelligence that provides machine learning-based suggestions on data wrangling. The graphical interface of Paxata is shown in Figure 3.8 given below, in which data append operation is performed on the particular column.

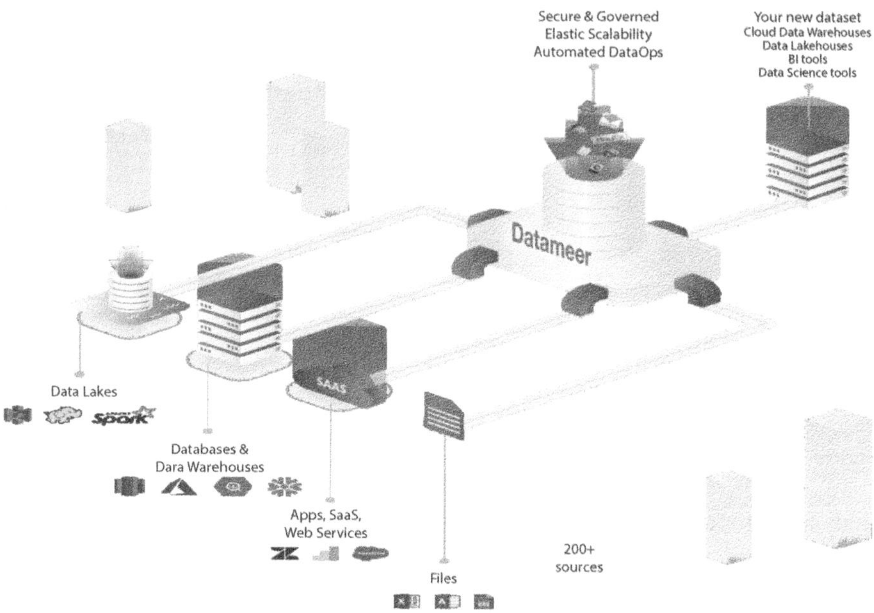

Figure 3.7 Graphical representation for accepting the input from various heterogeneous data sources and data files from web services and apps.

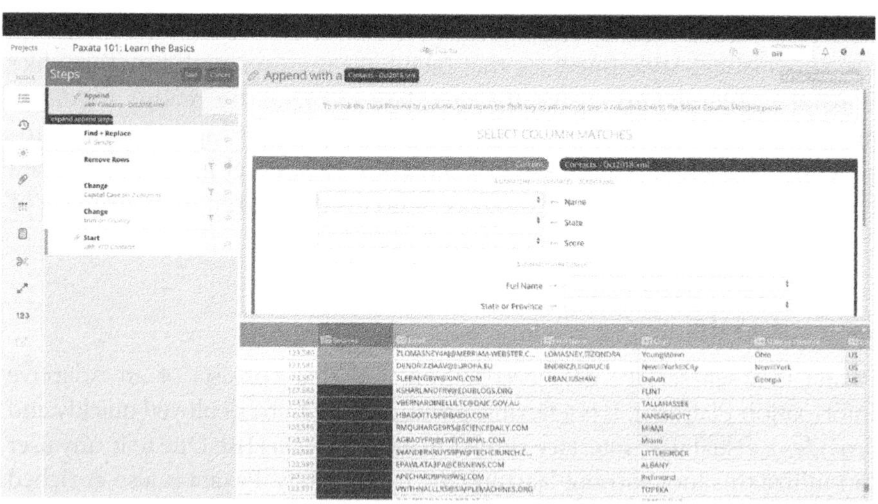

Figure 3.8 Image depicting the graphical user interface of Paxata tool performing the data append operation on a particular column.

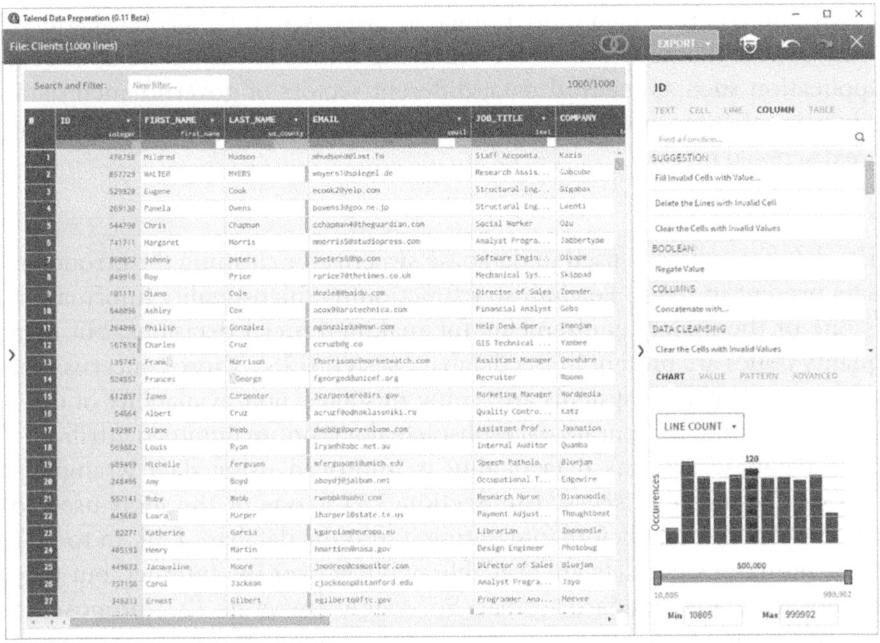

Figure 3.9 Image depicting data preparation process using Talend tool where suggestions are displayed according to columns in the dataset.

3.5.8 Talend

Talend is a data preparation and visualization tool used for data wrangling operations. It has a user-friendly and easy-to-use interface means non-technical people can use it for data preparation [15]. Machine learning-based algorithms have been deployed for data preparation operations such as cleaning, merging, transforming, and standardization. It is an automated product that provides the user with the suggestion at the time of data wrangling. The following Figure 3.9 depicts the data preparation process using Talend, in which recommendations have been displayed according to columns in the dataset.

3.6 Data Wrangling Application Areas

It has been observed that data wrangling is one of the initial and essential phase in any of the framework for the process in order to make the messy and complex data more unified as discussed in the earlier sections. Due to these characteristics, data wrangling is used in various fields of data application such as medical data, different sectors of governmental data, educational data, financial data, etc. Some of the significant applications are discussed below.

A. Database Systems
The data wrangling is used in database systems for cleaning the erroneous data present in them. For industry functioning, high-quality information is one of the major requirements for making crucial decisions, but data quality issues are present in the database systems[25]. Those concerns that exist in the database systems are typing mistakes, non-availability of data, redundant data, inaccurate data, obsolete data, not maintained attributes. Such database system's data quality is improved using data wrangling. Trifacta Wrangler (discussed in Section 3.5) is one of the tools used to pre-process the data before integrating it into the database [20]. In today's time, numerous datasets are available publicly over the internet, but they do not have any standard format. So, MacAveny *et al.* [22] proposed a robust and lightweight tool, ir_datasets, to manage the datasets (textual datasets) available over the internet. It provides the Python and command line-based interface for the users to explore the required information from the documents through ID.

B. Open government data
There is an availability of many open government data that can be brought into effective use, but extracting the usable data in the required form is a hefty task. Konstantinou *et al.* [2] proposed a data wrangling framework known as value-added data system (VADA). This architecture focuses on all the components of the data wrangling process, automating the process with the use of the available application domain information, using the user feedback for the refinement of results by considering the user's priorities. This proposed architecture is comparable to ETL and has been demonstrated on real estate data collected from web data and open government data specifying the properties for sale and areas for properties location respectively.

C. Traffic data
A number of domain-independent data wrangling tools have been constructed to overcome the problems of data quality in different applications. Sometimes, using generic data wrangling tools is a time-consuming process and also needs advanced IT skills for traffic analysts. One of the

shortcomings for the traffic datasets consisting of data generated from the road sensors is the presence of redundant records of the same moving object. This redundancy can be removed with the use of multiple attributes, such as device MAC address, vehicle identifier, time, and location of vehicle [21]. Another issue present in the traffic datasets is the missing data due to the malfunction of sensors or bad weather conditions affecting the proper functioning of sensors. This can be removed with the use of data with temporal or the same spatial characteristics.

D. Medical data

The datasets available in real time is heterogeneous data that contain artifacts. Such scenarios are mainly functional with the medical datasets as they have information from numerous resources, such as doctor's diagnosis, patient reports, monitoring sensors, etc. Therefore, to manage such dataset artifacts in medical datasets, Barrejón *et al.* [9] proposed the data wrangling tool using sequential variational autoencoders (VAEs) using the Shi-VAE methodology. This tool's performance is analyzed on the intensive care unit and passive human monitoring datasets based on root mean square error (RMSE) metrics. Ceusters *et al.* [23] worked on the ontological datasets proposing the technique based on referent tracking. In this, a template is presented for each dataset applied to each tuple in it, leading to the generation of referent tracking tuples created based on the unique identifier.

E. Journalism data

Journalism is one field where the journalist uses a lot of data and computations to report the news. To extract the relevant and accurate information, data wrangling is one of the journalist's significant tasks. Kasica *et al.* [24] have studied 50 publically available repositories and analysis code authored by 33 journalists. The authors have observed the extensive use of multiple tables in data wrangling on computational journalism. The framework is proposed for general mutitable data wrangling, which will support computational journalism and be used for general purposes.

In this section, the broad application areas have been explored, but the exploration can still not be made for the day-to-day wrangling processes.

3.7 Future Directions and Conclusion

In this technological era, having appropriate and accurate data is one of the prerequisites. To achieve this prerequisite, data analysts need to spend ample time producing quality data. Although data wrangling approaches are defined to achieve this target, data cleaning and integration are still one of the persistent issues present in the database community. This paper examines the basic terminology, challenges, architecture, tools available, and application areas of data wrangling.

Although the researchers highlighted the challenges, gaps, and potential solutions in the literature, there is still much room that can be explored in the future. There is a need to integrate the visual approaches with the existing techniques to extend the impact of the data wrangling process. The specification of the presence of errors and their fixation in the visual approaches should also be mentioned to better understand and edit operations through the user. The data analyst needs to be well expertise in the field of programming and the specific application area to utilize the relevant operations and tools for data wrangling to extract the meaningful insights of data.

References

1. Sutton, C., Hobson, T., Geddes, J., Caruana, R., Data diff: Interpretable, executable summaries of changes in distributions for data wrangling, in: *Proceedings of the 24th ACM SIGKDD International Conference on Knowledge Discovery & Data Mining*, pp. 2279–2288, 2018.
2. Konstantinou, N., Koehler, M., Abel, E., Civili, C., Neumayr, B., Sallinger, E., The VADA architecture for cost-effective data wrangling, in: *Proceedings of ACM International Conference on Management of Data*, pp. 1599–1602, 2017.
3. Bogatu, A., Paton, N.W., Fernandes, A.A., Towards automatic data format transformations: Data wrangling at scale, in: *British International Conference on Databases*, pp. 36–48, 2017.
4. Koehler, M., Bogatu, A., Civili, C., Konstantinou, N., Abel, E., Fernandes, A.A., Paton, N.W., Data context informed data wrangling, in: *2017 IEEE International Conference on Big Data (Big Data)*, pp. 956–963, 2017.
5. Furche, T., Gottlob, G., Libkin, L., Orsi, G., Paton, N.W., Data wrangling for big data: Challenges and opportunities, in: *EDBT*, vol. 16, pp. 473–478, 2016.
6. Kandel, S., Heer, J., Plaisant, C., Kennedy, J., Van Ham, F., Riche, N.H., Buono, P., Research directions in data wrangling: Visualizations and transformations for usable and credible data. *Inf. Vis.*, 10, 4, 271–288, 2011.
7. Braun, M.T., Kuljanin, G., DeShon, R.P., Special considerations for the acquisition and wrangling of big data. *Organ. Res. Methods*, 21, 3, 633–659, 2018.
8. Bors, C., Gschwandtner, T., Miksch, S., Capturing and visualizing provenance from data wrangling. *IEEE Comput. Graph. Appl.*, 39, 6, 61–75, 2019.

9. Barrejón, D., Olmos, P. M., Artés-Rodríguez, A., Medical data wrangling with sequential variational autoencoders. *IEEE J. Biomed. Health Inform.*, 2021.
10. Etaati, L., Data wrangling for predictive analysis, in: *Machine Learning with Microsoft Technologies*, Apress, Berkeley, CA, pp. 75–92, 2019.
11. Rattenbury, T., Hellerstein, J. M., Heer, J., Kandel, S., Carreras, C., *Principles of data wrangling: Practical techniques for data preparation*. O'Reilly Media, Inc., 2017.
12. Abedjan, Z., Morcos, J., Ilyas, I.F., Ouzzani, M., Papotti, P., Stonebraker, M., Dataxformer: A robust transformation discovery system, in: *2016 IEEE 32nd International Conference on Data Engineering (ICDE)*, pp. 1134–1145, 2016.
13. Datameer, Datameer spectrum, September 20, 2021. https://www.datameer.com/spectrum/.
14. Kosara, R., Trifacta wrangler for cleaning and reshaping data, September 29, 2021. https://eagereyes.org/blog/2015/trifacta-wrangler-for-cleaning-and-reshaping-data.
15. Zaharov, A., Datalytyx an overview of talend data preparation (beta), September29,2021.https://www.datalytyx.com/an-overview-of-talend-data-preparation-beta/.
16. Altair.com/Altair Monarch, Altair monarch self-service data preparation solution, September 29, 2021. https://www.altair.com/monarch.
17. Tabula.technology, Tabula: Extract tables from PDFs, September 29, 2021. https://tabula.technology/.
18. DataRobot | AI Cloud, Data preparation, September 29, 2021. https://www.paxata.com/self-service-data-prep/.
19. Cambridge Semantics, *Anzo Smart Data Lake 4.0-A Data Lake Platform for the Enterprise Information Fabric [Slideshare]*, September 29, 2021, https://www.cambridgesemantics.com/anzo-smart-data-lake-4-0-data-lake-platform-enterprise-information-fabric-slideshare/.
20. Azeroual, O., Data wrangling in database systems: Purging of dirty data. *Data*, 5, 2, 50, 2020.
21. Sampaio, S., Aljubairah, M., Permana, H.A., Sampaio, P.A., Conceptual approach for supporting traffic data wrangling tasks. *Comput. J.*, 62, 461–480, 2019.
22. MacAvaney, S., Yates, A., Feldman, S., Downey, D., Cohan, A., Goharian, N., Simplified data wrangling with ir_datasets, *Proceedings of the 44th International ACM SIGIR Conference on Research and Development in Information Retrieval*, 2429–2436, 2021.
23. Ceusters, W., Hsu, C.Y., Smith, B., Clinical data wrangling using ontological realism and referent tracking, in: *Proceedings of the Fifth International Conference on Biomedical Ontology (ICBO)*, pp. 27–32, 2014.
24. Kasica, S., Berret, C., Munzner, T., Table scraps: An actionable framework for multi-table data wrangling from an artifact study of computational journalism. *IEEE Trans. Vis. Comput. Graph.*, 27, 2, 957–966, 2020.

25. Swetha, K.R., Niranjanamurthy, M., Amulya, M.P., Manu, Y.M., Prediction of pneumonia using big data, deep learning and machine learning techniques. *2021 6th International Conference on Communication and Electronics Systems (ICCES)*, pp. 1697–1700, 2021, doi: 10.1109/ICCES51350.2021.9489188.

4

Essentials of Data Wrangling

Menal Dahiya, Nikita Malik* and Sakshi Rana

Dept. of Computer Applications, Maharaja Surajmal Institute, Janakpuri, New Delhi, India

Abstract
Fundamentally, data wrangling is an elaborate process of transforming, enriching, and mapping data from one raw data form into another, to make it more valuable for analysis and enhancing its quality. It is considered as a core task within every action that is performed in the workflow framework of data projects. Wrangling of data begins from accessing the data, followed by transforming it and profiling the transformed data. These wrangling tasks differ according to the types of transformations used. Sometimes, data wrangling can resemble traditional extraction, transformation, and loading (ETL) processes. Through this chapter, various kinds of data wrangling and how data wrangling actions differ across the workflow are described. The dynamics of data wrangling, core transformation and profiling tasks are also explored. This is followed by a case study based on a dataset on forest fires, modified using Excel or Python language, performing the desired transformation and profiling, and presenting statistical and visualization analyses.

Keywords: Data wrangling, workflow framework, data transformation, profiling, core profiling

4.1 Introduction

Data wrangling, which is also known as data munging, is a term that involves mapping data fields in a dataset starting from the source (its original raw form) to destination (more digestible format). Basically, it consists of variety of tasks that are involved in preparing the data for further analysis. The methods that you will apply for wrangling the data totally

**Corresponding author*: nikitamalik@msijanakpuri.com

M. Niranjanamurthy, Kavita Sheoran, Geetika Dhand, and Prabhjot Kaur (eds.) Data Wrangling: Concepts, Applications and Tools, (71–90) © 2023 Scrivener Publishing LLC

depends on the data that you are working on and the goal you want to achieve. These methods may differ from project to project. A data wrangling example could be targeting a field, row, or column in a dataset and implementing an action like cleaning, joining, consolidating, parsing or filtering to generate the required output. It can be a manual or machine-driven process. In cases where datasets are exceptionally big, automated data cleaning is required.

Data wrangling term is derived and defined as a process to prepare the data for analysis with data visualization aids that accelerates the faster process [1]. If the data is accurately wrangled then it ensures that we have entered quality data into analytics process. Data wrangling leads to effective decision making. Sometimes, for making any kind of required manipulation in the data infrastructure, it is necessary to have appropriate permission.

During the past 20 years, processing on data and the urbanity of tools has progressed, which makes it more necessary to determine a common set of techniques. The increased availability of data (both structured and unstructured) and the utter volume of it that can be stored and analyzed has changed the possibilities for data analysis—many difficult questions are now easier to answer, and some previously impossible ones are within reach [2]. There is a need for glue that helps to tie together the various parts of the data ecosystem, from JSON APIs (JavaScript Object Notation Application Programming Interface) to filtering and cleaning data to creating understandable charts. In addition to classic typical data, quality criteria such as accuracy, completeness, correctness, reliability, consistency, timeliness, precision, and conciseness are also an important aspect [3].

Some tasks of data wrangling include:

1. Creating a dataset by getting data from various data sources and merging them for drawing the insights from the data.
2. Identifying the outliers in the given dataset and eliminating them by imputing or deleting them.
3. Removal of data that is either unnecessary or irrelevant to the project.
4. Plotting graphs to study the relationship between the variables and to identify the trend and patterns across.

4.2 Holistic Workflow Framework for Data Projects

This section presents a framework that shows how to work with data. As one moves through the process of accessing, transforming, and using the data, there are certain common sequences of actions that are performed. The goal is to cover each of these processes. Data wrangling also constitutes a promising direction for visual analytics research, as it requires combining automated techniques (example, discrepancy detection, entity resolution, and semantic data type inference) with interactive visual interfaces [4].

Before deriving direct, automated value we practice to derive indirect, human-mediated value from the given data. For getting the expected valuable result by an automated system, we need to assess whether the core quality of our data is sufficient or not. Report generation and then analyzing it is a good practice to understand the wider potential of the data. Automated systems can be designed to use this data.

This is how the data projects progress: starting from short-term answering of familiar questions, to long-term analyses that assess the quality and potential applications of a dataset and at last to designing the systems that will use the dataset in an automated way. Undergoing this complete process our data moves through three main stages of data wrangling: raw, refined, and production, as shown in Table 4.1.

4.2.1 Raw Stage

Discovering is the first step to data wrangling. Therefore, in the raw stage, the primary goal is to understand the data and getting an overview of your data. To discover what kinds of records are in the data, how are the record fields encoded and how does the data relate to your organization, to the kinds of operations you have, and to the other existing data you are using. Get familiar with your data.

Table 4.1 Movement of data through various stages.

Primary objectives	**Data stage**		
	Raw	**Refined**	**Production**
	• Source data as it is, with no transformation, ingest data • Discovering the data and creation of metadata	• Data is discovered, explored and experimented for hypothesis validation and tests. • Data cleaning, Conduct analyses, intense exploration and forecasting.	• Creation of production-quality data. • Clean and well-structured data is stored in the optimal format.

4.2.2 Refined Stage

After seeing the trends and patterns that will be helpful to conceptualize what kind of analysis you may want to do and being armed with an understanding of the data, you can then refine the data for intense exploration. When you collect raw data, initially are in different sizes and shapes, and do not have any definite structure. We can remove parts of the data that are not being used, reshaping the elements that are poorly formatted, and establishing relationships between multiple datasets. Data cleaning tools are used to remove errors that could influence your downstream analysis in a negative manner.

4.2.3 Production Stage

Once the data to be worked with is properly transformed and cleaned for analysis after completely understanding it, it is time to decide if all the data needed for the task at hand is there. Once the quality of data and its potential applications in automated systems are understood, the data can be moved to the next stage, that is, the production stage. On reaching this point, the final output is pushed downstream for the analytical needs.

Only a minority of data projects ends up in the raw or production stages, and the majority end up in the refined stage. Projects ending in the refined stage will add indirect value by delivering insights and models that drive better decisions. In some cases, these projects might last multiple years [2].

4.3 The Actions in Holistic Workflow Framework

4.3.1 Raw Data Stage Actions

There are mainly three actions that we perform in the raw data stage as shown in Figure 4.1.

- Focused on outputting data, there are two ingestion actions:
  1. Ingestion of data
- Focused on outputting insights and information derived from the data:
  2. Creating the generic metadata
  3. Creating the propriety metadata.

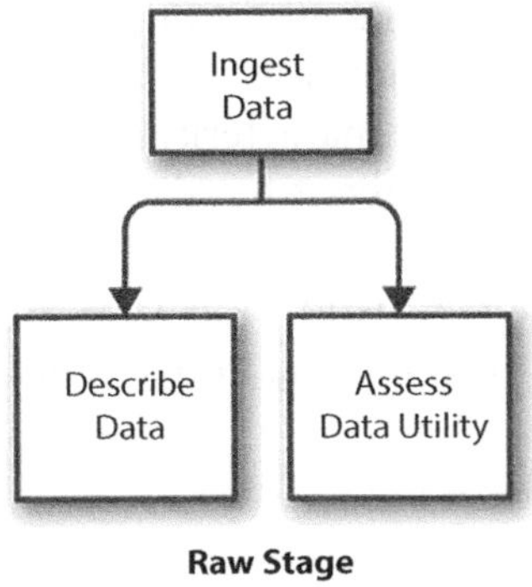

Figure 4.1 Actions performed in the raw data stage.

4.3.1.1 *Data Ingestion*

Data ingestion is the shipment of data from variegated sources to a storage medium where it can be retrieved, utilized, and analyzed to a data warehouse, data mart or database. This is the key step for analytics. Because of the various kinds of spectrum, the process of ingestion can be complex in some areas. In less complex areas many persons get their data as files through channels like FTP websites, emails. Other more complex areas include modern open-source tools which permit more comminuted and real-time transfer of data. In between these, more complex and less complex spectrum are propriety platforms, which support a variety of data transfer. These include Informatica Cloud, Talend, which is easy to maintain even for the people who does not belong to technical areas.

In the traditional enterprise data warehouses, some initial data transformation operations are involved in ingestion process. After the transformation when it is totally matched to the syntaxes that are defined by the warehouse, the data is stored in locations which are predefined. In some cases, we have to add on new data to the previous data. This process of appending newly arrived data can be complex if the new data contains edit to the previous data. This leads to ingest new data into separate locations, where certain rules can be applied for merging during the process of refining. In some areas, it can be simple where we just add new records at the end of the prior records.

4.3.1.2 *Creating Metadata*

This stage occurs when the data that you are ingesting is unknown. In this case, you do not how to work with your data and what results can you expect from it. This leads to the actions that are related to the creation of metadata. One action is known as creating generic metadata, which focuses on understanding the characteristics of your data. Other action is of making a determination about the data's value by using the characteristics of your data. In this action, custom metadata is created. Dataset contains records and fields, which means rows and columns. You should focus on understanding the following things while describing your data:

- Structure
- Accuracy
- Temporality
- Granularity
- Scope of your data

Based on the potential of your present data, sometimes, it is required to create custom metadata in the discovery process. Generic metadata is useful to know how to properly work with the dataset, whereas custom metadata is required to perform specific analysis.

4.3.2 Refined Data Stage Actions

After the ingestion and complete understanding of your raw data, the next essential step includes the refining of data and exploring the data through analyses. Figure 4.2 shows the actions performed in this stage.

The primary actions involve in this stage are:

- Responsible for generating refined data which allows quick application to a wide range of analyses:
  1. Generate Ad-Hoc Reports
- Responsible for generating insights and information that are generated from the present data, which ranges from general reporting to further complex structures and forecasts:
  2. Prototype modeling

The all-embracing motive in designing and creating the refined data is to simplify the predicted analyses that have to perform. As we will not foresee all of the analyses that have to be performed; therefore, we look at the patterns that are derived from the initial analyses, draw insights and get inspired from them to create new analysis directions that we had not considered previously. After refining the datasets, we compile them or modify them. Very often, it is required to repeat the actions in refining stage.

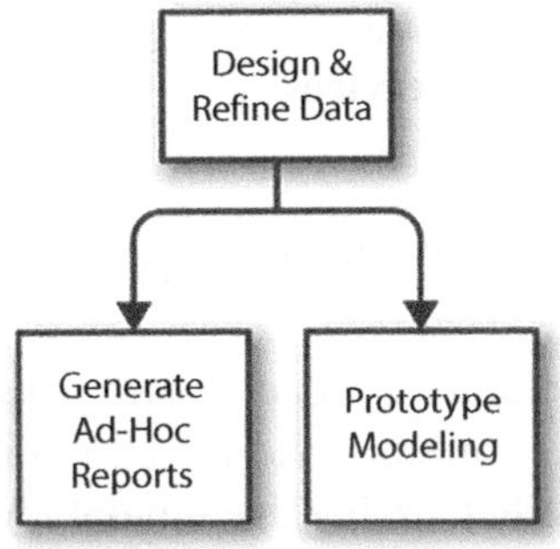

Figure 4.2 Actions performed in refined data stage.

In this stage, our data is transformed the most in the process of designing and preparing the refined data. While creating the metadata in the raw stage if there we any errors in the dataset's accuracy, time, granularity, structure or scope, those issues must be resolved here during this stage.

4.3.3 Production Data Stage Actions

After refining the data, we reach at a stage where we start getting valuable insights from the dataset, its time separating the analyses (Figure 4.3). By separating, it means that you will be able to detect which analyses you have to do on a regular basis and which ones were enough for one-time analyses.

- Even after refining the data, when creating the production data, it is required to optimize your data. After that monitoring and scheduling the flow of this ideal data after

Figure 4.3 Actions performed in production data stage.

optimization and maintaining regular reports and data-driven products and services.

4.4 Transformation Tasks Involved in Data Wrangling

Data wrangling is a core iterative process that throws up the cleanest, most useful data possible before you start your actual analysis [5]. Transformation is one of the core actions that are involved in data wrangling. Another task is profiling, and we need to quick iterate between these two actions. Now we will explore the transformation tasks that are present in the process of data wrangling.

These are the core transformation actions that we need to apply on the data:

- Structuring
- Enriching
- Cleansing

4.4.1 Structuring

These are the actions that are used to change the schema and form of the data. Structuring mainly involves shifting records around and organizing the data. It is a very simple kind of transformation; sometimes it can be just changing the order of columns within a table. It also includes summarizing record field values. In some cases, it is required to break record fields into subcomponents or combining fields together which results in a complex transformation.

The most complex kind of transformation is the inter-record structuring which includes aggregations and pivots of the data:

Aggregation—It allows switching in the granularity of the dataset. For example, switching from individual person to segment of persons.

Pivoting—It includes shifting entries (records) into columns (fields) and vice-versa.

4.4.2 Enriching

These are the actions that are used to add elementary new records from multiple datasets to the dataset and strategize about how this new additional data might raise it. The typical structuring transformations are:

- ➢ Join: It combines data from various tables based on a matching condition between the linking records.
- ➢ Union: It combines the data into new rows by blending multiple datasets together. It concatenates rows from different datasets by matching up rows It returns distinct rows.

Besides joins and unions, insertion of metadata and computing new data entries from the existing data in your dataset which results in the generation of generic metadata is another common task. This inserted metadata can be of two types:

- Independent of the dataset
- Specific to the dataset

4.4.3 Cleansing

These are the actions that are used to resolve the errors or to fix any kind of irregularities if present in your dataset. It fixes the quality and consistency issues and makes the dataset clean. High data quality is not just desirable, but one of the main criteria that determine whether the project is successful, and the resulting information is correct [6]. It basically includes manipulating single column values within the rows. The most common type is to fix the missing or the NULL values in the dataset, implementing formatting and hence increasing the quality of data.

4.5 Description of Two Types of Core Profiling

In order to understand your data before you start transforming or analyzing it, the first step is profiling. Profiling leads to data transformations. This helps in reviewing source data for content and better quality [7].

One challenge of data wrangling is that reformatting and validating data require transforms that can be difficult to specify and evaluate. For instance, splitting data into meaningful records and attributes often involves regular expressions that are error-prone and tedious to interpret [8, 9].

Profiling can be divided on the basis of unit of data they work on. There are two kinds of profiling:

- Individual values profiling
- Set-based profiling

4.5.1 Individual Values Profiling

There are two kinds of constraints in individual values profiling. These are:

1. Syntactic
2. Semantic

4.5.1.1 Syntactic

It focuses on the formats, for example, the format of date is MM-DD-YYYY. Therefore, date value should be in this format only.

4.5.1.2 Semantic

Semantic constraints are built in context or exclusive business logic; for example, your company is closed for business on a festival so no transactions should exist on that particular day. This helps us to determine if the individual record field value or entire record is valid or not.

4.5.2 Set-Based Profiling

This kind of profiling mainly focuses on the shape of values and how the data is distributed within a single record field or in the range of relationships between more than one record field.

For example, there might be higher retail sales in holidays than a non-holiday. Thus, you could build a set-based profile to ensure that sales are distributed across the month as it was expected.

4.6 Case Study

Wrangle the data into a dataset that provides meaningful insights to carry-out cleansing process; it requires writing codes in idiosyncratic characters in languages of Perl, R and editing manually with tools like MS-Excel [10].

- In this case study, we have a Brazilian Fire Dataset, as shown in Figure 4.4 (https://product2.s3-ap-southeast-2.amazonaws.com/Activity_files/MC_DAP01/Brazilian-fire-dataset.csv). The goal is to perform the following tasks:
  - Interpretation of the imported data through a dataset
  - Descriptive statistics of the dataset

Figure 4.4 This is how the dataset looks like. It consists of number of records in it.

- Plotting graphs
- Creating a Data Frame and working on certain activities using Python

Kandel *et al.* [11] have discussed a wide range of topics and problems in the field of data wrangling, especially with regard to visualization. For example, graphs and charts can help identify data quality issues, such as missing values.

4.6.1 Importing Required Libraries

- Pandas, NumPy, and Matplotlib
- Pandas is a Python library for data analysis. Padas is built on top of two core Python libraries—matplotlib for data visualization and NumPy for mathematical operations.
- How we import these libraries can be seen in Figure 4.5 below

In this above code, we created a DataFrame by the name of df_fire and in this DataFrame we have loaded a csv file using Pandas read_csv( )

Figure 4.5 Snippet of libraries included in the code.

```
In [3]: df_fire = pd.read_csv('brazilian-fire-dataset.csv')
        df_fire
Out[3]:
```

	Year	State	Month	Number of Fires	Date Reported
0	1998	Acre	January	0.0	1/01/1998
1	1999	Acre	January	0.0	1/01/1999
2	2000	Acre	January	0.0	1/01/2000
3	2001	Acre	January	0.0	1/01/2001
4	2002	Acre	January	0.0	1/01/2002
...	...	...	...	...	...
6449	2012	Tocantins	December	128.0	1/01/2012
6450	2013	Tocantins	December	85.0	1/01/2013
6451	2014	Tocantins	December	223.0	1/01/2014
6452	2015	Tocantins	December	373.0	1/01/2015
6453	2016	Tocantins	December	119.0	1/01/2016

6454 rows × 5 columns

Figure 4.6 Snippet of dataset used.

function. Full Path and Name of the file is 'brazilian-fire-dataset.csv'. The result is shown in Figure 4.6.

Here we can see that the total number of records is 6454 rows and there are five columns. The column "Number of Fires" is having float datatype.

4.6.2 Changing the Order of the Columns in the Dataset

In the first line of code, we are specifying the order of the column. In second line we have changed the datatype of column "Number of Fires" to Integer type. Then we will rearrange the columns in the dataset and print it. The result is shown in Figure 4.7 and Figure 4.8.

4.6.3 To Display the DataFrame (Top 10 Rows) and Verify that the Columns are in Order

For displaying top 10 records of the dataset the .head() function is used as follows (Figure 4.9).

```
In [4]: order_col = [ 'Number of Fires', 'State','Date Reported', 'Month', 'Year']
        df_fire['Number of Fires'] = df_fire['Number of Fires'].astype(int)
        df_fire = df_fire[order_col]
        df_fire
```

Figure 4.7 Snippet of manipulations on dataset.

Out[4]:

	Number of Fires	State	Date Reported	Month	Year
0	0	Acre	1/01/1998	January	1998
1	0	Acre	1/01/1999	January	1999
2	0	Acre	1/01/2000	January	2000
3	0	Acre	1/01/2001	January	2001
4	0	Acre	1/01/2002	January	2002
...	...	...	...	...	...
6449	128	Tocantins	1/01/2012	December	2012
6450	85	Tocantins	1/01/2013	December	2013
6451	223	Tocantins	1/01/2014	December	2014
6452	373	Tocantins	1/01/2015	December	2015
6453	119	Tocantins	1/01/2016	December	2016

6454 rows × 5 columns

Figure 4.8 The order of the columns has been changed and the datatype of "Number of fires" has been changed from float to int.

```
In [5]: df_fire.head(10)
```

Out[5]:

	Number of Fires	State	Date Reported	Month	Year
0	0	Acre	1/01/1998	January	1998
1	0	Acre	1/01/1999	January	1999
2	0	Acre	1/01/2000	January	2000
3	0	Acre	1/01/2001	January	2001
4	0	Acre	1/01/2002	January	2002
5	10	Acre	1/01/2003	January	2003
6	0	Acre	1/01/2004	January	2004
7	12	Acre	1/01/2005	January	2005
8	4	Acre	1/01/2006	January	2006
9	0	Acre	1/01/2007	January	2007

Figure 4.9 Top 10 records of the dataset.

4.6.4 To Display the DataFrame (Bottom 10 rows) and Verify that the Columns Are in Order

For displaying top 10 records of the dataset, we use .tail( ) function as follows (Figure 4.10).

4.6.5 Generate the Statistical Summary of the DataFrame for All the Columns

To get the statistical summary of the data frame for all the columns we use the .describe() function. The result is shown in Figure 4.11.

```
In [6]: df_fire.tail(10)
```

Out[6]:

	Number of Fires	State	Date Reported	Month	Year
6444	13	Tocantins	1/01/2007	December	2007
6445	7	Tocantins	1/01/2008	December	2008
6446	46	Tocantins	1/01/2009	December	2009
6447	72	Tocantins	1/01/2010	December	2010
6448	105	Tocantins	1/01/2011	December	2011
6449	128	Tocantins	1/01/2012	December	2012
6450	85	Tocantins	1/01/2013	December	2013
6451	223	Tocantins	1/01/2014	December	2014
6452	373	Tocantins	1/01/2015	December	2015
6453	119	Tocantins	1/01/2016	December	2016

Figure 4.10 Result—Bottom 10 records of the dataset.

```
In [7]: df_fire.describe(include='all')
```

Out[7]:

	Number of Fires	State	Date Reported	Month	Year
count	6454.000000	6454	6454	6454	6454.000000
unique	NaN	27	20	12	NaN
top	NaN	Alagoas	1/01/2001	January	NaN
freq	NaN	240	324	541	NaN
mean	108.235358	NaN	NaN	NaN	2007.461729
std	190.843947	NaN	NaN	NaN	5.746654
min	0.000000	NaN	NaN	NaN	1998.000000
25%	3.000000	NaN	NaN	NaN	2002.000000
50%	24.000000	NaN	NaN	NaN	2007.000000
75%	113.000000	NaN	NaN	NaN	2012.000000
max	998.000000	NaN	NaN	NaN	2017.000000

Figure 4.11 Here we can get the count, unique, top, freq, mean, std, min, quartiles & percentiles, max etc. of all the respected columns.

4.7 Quantitative Analysis

4.7.1 Maximum Number of Fires on Any Given Day

Here, first we will get the maximum number of fires on any given day in the dataset by using the .max( ) function. Then we will display the record that is having this number of fires. The result is shown in Figure 4.12.

```
In [8]: max_no_fire = df_fire['Number of Fires'].max()
        df_fire[df_fire['Number of Fires']==max_no_fire]
```

Out[8]:

	Number of Fires	State	Date Reported	Month	Year
888	998	Amazonas	1/01/2008	September	2008

```
In [ ]: #ANSWER- Maximum number of Fires (998 fires) were reported in the month of September 2008 in the state of Amazonas
```

Figure 4.12 Maximum number of fires is 998 and was reported in the month of September 2000 in the state of Amazonas.

4.7.2 Total Number of Fires for the Entire Duration for Every State

- Pandas group by is used for grouping the data according to the categories and apply a function to the categories. It also helps to aggregate data efficiently. Pandas dataframe.groupby() function is used to split the data into groups based on some criteria. Pandas objects can be split on any of their axes. The abstract definition of grouping is to provide a mapping of labels to group names [12].
- .agg( ) Dataframe.aggregate() function is used to apply some aggregation across one or more column. Aggregate using callable, string, dict, or list of string/callables. Most frequently used aggregations are: sum, min and max [13, 14].

The result is shown in Figure 4.13 below.

For example, Acre-18452, Bahia-44718 etc. Here because of the .head() function we are able to see only top 10 values.

```
In [9]: state_fires = df_fire.groupby(['State']).agg({'Number of Fires': np.sum})
        state_fires.head()
```

Out[9]:

State	Number of Fires
Acre	18452
Alagoas	4644
Amapá	21831
Amazonas	30636
Bahia	44718

Figure 4.13 The data if grouped by state and we can get the total number of fires in a particular state.

```
In [10]: state_fires.describe()
Out[10]:
```

	Number of Fires
count	27.000000
mean	25872.259259
std	14251.935889
min	3237.000000
25%	15619.500000
50%	24459.000000
75%	37565.000000
max	51118.000000

```
In [11]: print("Max of Total Fires recorded were %d, and this was for state: %s"\
             %(state_fires['Number of Fires'].max(),\
                   state_fires.idxmax()[0]))

         print("Min of Total Fires recorded were %d, and this was for state: %s"\
             %(state_fires['Number of Fires'].min(),\
                   state_fires.idxmin()[0]))

         Max of Total Fires recorded were 51118, and this was for state: São Paulo
         Min of Total Fires recorded were 3237, and this was for state: Sergipe
```

Figure 4.14 Maximum of total fires recorded were 51118, and this was for State—Sao Paulo Minimum of total fires recorded were 3237, and this was for State—Sergipe.

4.7.3 Summary Statistics

- By using .describe() we can get the statistical summary of the dataset (Figure 4.14).

4.8 Graphical Representation

4.8.1 Line Graph

Following in Figure 4.15 code is given. Here Plot function in matplotlib is used.

In Figure 4.16, the line plots depict the values on the series of data points that are connected with straight lines.

4.8.2 Pie Chart

For getting the values of total numbers of fires in a particular month, we will again use the GroupBy and aggregate function and get the month fires.

```
In [12]: #Line graph Number of Fires vs Record Number
         fig=plt.figure(figsize=(15,6))
         plt.plot(df_fire['Number of Fires'], color='purple')
         plt.xlabel("Record Number",fontsize=16)
         plt.ylabel("Number of Fires",fontsize=16)
         plt.title("Line graph Number of Fires vs Record Number",fontsize=20)
         plt.show()
```

Figure 4.15 Code snippet for line graph.

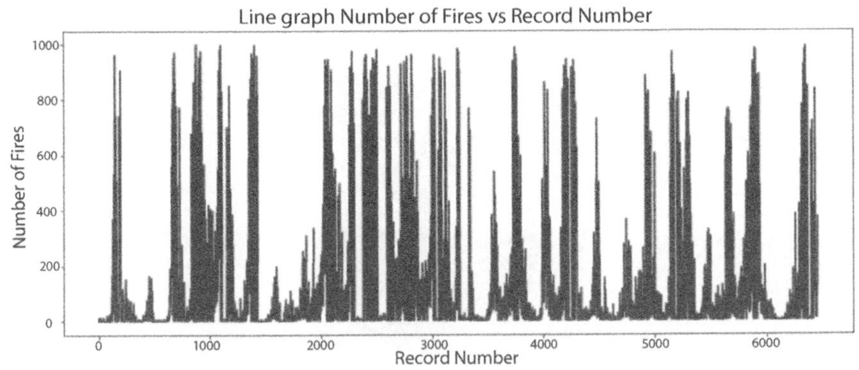

Figure 4.16 Line graph.

```
In [13]: #Pie Chart for Number of Fires in a particular Month
         month_fires = df_fire.groupby(['Month']).agg({'Number of Fires': np.sum})
         month_fires
```

Out[13]:

Month	Number of Fires
April	28188
August	87970
December	57513
February	30848
January	47746
July	92302
June	56003
March	30717
May	34729
November	85471
October	88600
September	58464

Figure 4.17 Code snippet for creating pie graph.

After getting the required data, we will plot the pie chart as given in Figure 4.18.

In Figure 4.18, we can see that the months of July, October, and November are having the highest numbers of fires. It is showing percentages of a whole, and it represents percentages at a set point in time. Pie charts do not show changes over time.

4.8.3 Bar Graph

For plotting the bar graph, we have to get the values for the total number of fires in a particular year (Figure 4.19).

```
In [14]: m_values=(47746,30848,30717,28188,34729,56003,92301,87978,58464,58600,85471,57513)
         m_labels=('January', 'February','March','April','May','June','July', 'August','Septemer','October','November','December')
         fig=plt.figure(figsize=(15,6))
         plt.pie(m_values,labels=m_labels)
         plt.title('Pie Chart for Number of Fires in a particular Month', fontsize=16)
         plt.show()
```

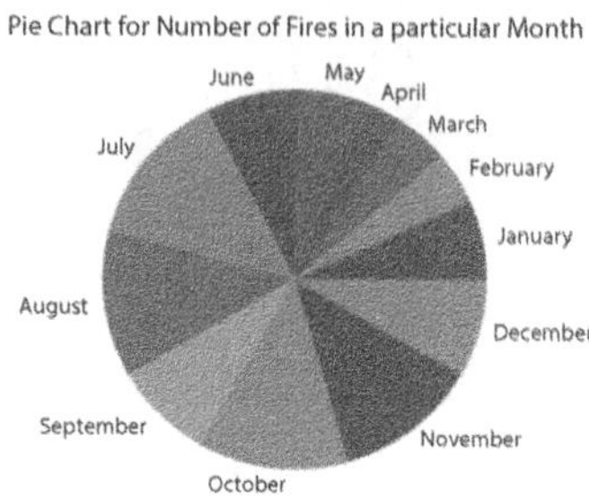

Figure 4.18 Pie chart.

```
In [15]: #Bar Graph Year VS Number of Fires in Descending order
         year_fires = df_fire.groupby(['Year']).agg({'Number of Fires': np.sum})
         year_fires.reset_index(level=0, inplace=True)
         year_fires.sort_values(by="Number of Fires", ascending=False, inplace=True)
         year_fires
```

Out[15]:

	Year	Number of Fires
5	2003	42737
18	2016	42189
17	2015	41184
14	2012	40065
16	2014	39601
11	2009	39104
6	2004	38427
4	2002	37369
12	2010	37015
19	2017	36663
15	2013	35130
7	2005	34984
13	2011	34619
8	2006	33819
9	2007	33017
10	2008	29363
3	2001	29056
2	2000	27341
1	1999	26869
0	1998	19999

Figure 4.19 Code snippet for creating bar graph.

```
In [16]: y_values=(42737,42189,41184,40065,39601,39104,38427,37369,37015,36663,35130,34984,34619,33819,33017,29363,29056,27421,26869,1999
         y_labels=('2003','2016','2015','2012','2014','2009','2004','2002','2010','2017','2013','2005','2011','2006','2007','2008','2001',
         fig=plt.figure(figsize=(15,6))
         plt.bar(y_labels,y_values,linestyle='dashed')
         plt.xlabel("Year", fontsize=16)
         plt.ylabel("Count of the Fires", fontsize=16)
         plt.title("Bar Graph Year VS Number of Fires in Descending order", fontsize=20)
         plt.show()
```

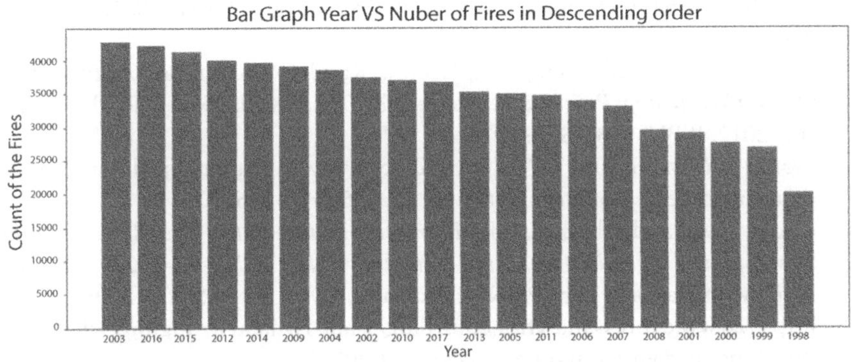

Figure 4.20 Bar graph.

After getting the values of the year and the number of fires in descending order, we will plot the bar graph. We use bar function from Matplotlib to achieve it (Figure 4.20).

In Figure 4.20, it can be observed that the highest number of fires is in the year 2003 and the lowest is in 1998. The graph shows the number of fires in decreasing order.

4.9 Conclusion

With the increasing rate of data amount and vast quantities of diverse data sources providing this data, many issues are faced by organizations. They are being compelled to use the available data and to produce competitive benefits for pulling through in the long run. For this, data wrangling offers an apt solution, of which, data quality is a significant aspect. Actions in data wrangling can further be divided into three parts which describe how the data is progressed through different stages. Transformation and profiling are the core processes which help us to iterate through records, add new values and, to detect errors and eliminate them. Data wrangling tools also help us to discover the problems present in data such as outliers, if any. Many quality problems can be recognized by inspecting the raw data; others can be detected by diagrams or other various kinds of representations. Missing values, for instance, are indicated by gaps in the graphs, wherein the type of representation plays a crucial role as it has great influence.

References

1. Cline, D., Yueh, S., Chapman, B., Stankov, B., Gasiewski, A., Masters, D., Mahrt, L., NASA cold land processes experiment (CLPX 2002/03): Airborne remote sensing. *J. Hydrometeorol.*, United States of America, 10, 1, 338–346, 2009.
2. Rattenbury, T., Hellerstein, J.M., Heer, J., Kandel, S., Carreras, C., *Principles of Data Wrangling: Practical Techniques for Data Preparation*, O'Reilly Media, Inc, 2017. ISBN: 9781491938928
3. Wang, R.Y. and Strong, D.M., Beyond accuracy: What data quality means to data consumers. *J. Manage. Inf. Syst.*, 12, 4, 5–33, 1996.
4. Cook, K.A. and Thomas, J.J., *Illuminating the Path: The Research and Development Agenda for Visual Analytics (No. PNNL-SA-45230)*, Pacific Northwest National Lab (PNNL), Richland, WA, United States, 2005.
5. https://www.expressanalytics.com/blog/what-is-data-wrangling-what-are-the-steps-in-data-wrangling/ [Date: 2/4/2022]
6. Rud, O.P., *Data Mining Cookbook: Modeling Data for Marketing, Risk, and Customer Relationship Management*, John Wiley & Sons, United States of America and Canada, 2001. ISBN-10 0471385646
7. https://panoply.io/analytics-stack-guide/ [Date: 2/5/2022]
8. Blackwell, A.F., XIII SWYN: A visual representation for regular expressions, in: *Your Wish is My Command*, pp. 245–270, Morgan Kaufmann, Massachusetts, United States of America, 2001. ISBN: 9780080521459
9. Scaffidi, C., Myers, B., Shaw, M., Intelligently creating and recommending reusable reformatting rules, in: *Proceedings of the 14th International Conference on Intelligent User Interfaces*, pp. 297–306, February 2009.
10. Kandel, S., Heer, J., Plaisant, C., Kennedy, J., Van Ham, F., Riche, N.H., Buono, P., Research directions in data wrangling: Visualizations and transformations for usable and credible data. *Inf. Vis.*, 10, 4, 271–288, 2011.
11. Kandel, S., Heer, J., Plaisant, C., Kennedy, J., Van Ham, F., Riche, N.H., Buono, P., Research directions in data wrangling: Visualizations and transformations for usable and credible data. *Inf. Vis.*, 10, 4, 271–288, 2011.
12. https://www.geeksforgeeks.org/python-pandas-dataframe-groupby/ Date: 03/05/2022]
13. https://www.geeksforgeeks.org/python-pandas-dataframe-aggregate/ [Date: 12/11/2021].
14. Swetha, K.R., Niranjanamurthy, M., Amulya, M.P., Manu, Y.M., Prediction of pneumonia using big data, deep learning and machine learning techniques. *2021 6th International Conference on Communication and Electronics Systems (ICCES)*, pp. 1697–1700, 2021, doi: 10.1109/ICCES51350.2021.9489188.

5

Data Leakage and Data Wrangling in Machine Learning for Medical Treatment

P.T. Jamuna Devi[1]* and B.R. Kavitha[2]

[1]*J.K.K. Nataraja College of Arts and Science, Komarapalayam, Tamilnadu, India*
[2]*Vivekanandha College of Arts and Science, Elayampalayam, Tamilnadu, India*

Abstract
Currently, healthcare and life sciences overall have produced huge amounts of real-time data by ERP (enterprise resource planning). This huge amount of data is a tough task to manage, and intimidation of data leakage by inside workers increases, the companies are wiping far-out for security like digital rights management (DRM) and data loss prevention (DLP) to avert data leakage. Consequently, data leakage system also becomes diverse and challenging to prevent data leakage. Machine learning methods are utilized for processing important data by developing algorithms and a set of rules to offer the prerequisite outcomes to the employees. Deep learning has an automated feature extraction that holds the vital features required for problem solving. It decreases the problem of the employees to choose items explicitly to resolve the problems for unsupervised, semisupervised, and supervised healthcare data. Finding data leakage in advance and rectifying for it is an essential part of enhancing the definition of a machine learning problem. Various methods of leakage are sophisticated and are best identified by attempting to extract features and train modern algorithms on the problem. Data wrangling and data leakage are being handled to identify and avoid additional processes in healthcare in the immediate future.

***Keywords*:** Data loss prevention, data wrangling, digital rights management, enterprise resource planning, data leakage

5.1 Introduction

Currently, in enterprise resource planning (ERP) machine learning and deep learning perform an important role. In the practice of developing

**Corresponding author*: jamunadevimphil@gmail.com

M. Niranjanamurthy, Kavita Sheoran, Geetika Dhand and Prabhjot Kaur (eds.) Data Wrangling: Concepts, Applications and Tools, (91–108) © 2023 Scrivener Publishing LLC the analytical model with machine learning or deep learning the data set is gathered as of several sources like a sensor, database, file, and so on [1]. The received data could not be utilized openly to perform the analytical process. To resolve this dilemma two techniques such as data wrangling and data preprocessing are used to perform Data Preparation [2]. An essential part of data science is data preparation. It is made up of two concepts like feature engineering and data cleaning. These two are inevitable to obtain greater accuracy and efficiency in deep learning and machine learning tasks [3]. Raw information is transformed into a clean data set by using a procedure is called data preprocessing. Also, each time data is gathered from various sources in raw form which is not sustainable for the analysis [4]. Hence, particular stages are carried out to translate data into a tiny clean dataset. This method is implemented in the previous implementation of Iterative Analysis. The sequence of steps is termed data preprocessing. It encompasses data cleaning, data integration, data transformation, and data reduction.

At the moment of creating an interactive model, the Data Wrangling method is performed. In other terms, for data utilization, it is utilized to translate raw information into a suitable format. This method is also termed Data Munging. This technique also complies with specific steps like subsequently mining the data from various sources, the specific algorithm is performed to sort the data, break down the data into a dispersed structured form, and then finally the data is stored into a different database [5]. To attain improved outcomes from the applied model in deep learning and machine learning tasks the data structure has to be in an appropriate way. Some specific deep learning and machine learning type require data in a certain form, for instance, null values are not supported by the Random Forest algorithm, and thus to carry out the random forest algorithm null values must be handled from the initial raw data set [6]. An additional characteristic is that the dataset needs to be formatted in such a manner that there are more than single deep learning and machine learning algorithm is performed in the single dataset and the most out of them has been selected. Data wrangling is an essential consideration to implement the model. Consequently, data is transformed to the appropriate possible format prior to utilizing any model intro it [7]. By executing, grouping, filtering, and choosing the correct data for the precision and implementation of the model might be improved. An additional idea is that once time series data must be managed each algorithm is performed with various characteristics. Thus, the time series data is transformed into the necessary structure of the applied model by utilizing Data Wrangling [8]. Consequently, the complicated data is turned into a useful structure for carrying out an evaluation.

5.2 Data Wrangling and Data Leakage

Data wrangling is the procedure of cleansing and combining complex and messy data sets for simple access and evaluation. With the amount of data and data sources fast-growing and developing, it is becoming more and more important for huge amounts of available data to be organized for analysis. Such a process usually comprises manually transforming and mapping data from a single raw form into a different format to let for more practical use and data organization. Deep learning and machine learning perform an essential role in the modern-day enterprise resource planning (ERP). In the practice of constructing the analytical model with machine learning or deep learning the data set is gathered as of a variety of sources like a database, file, sensors, and much more. The information received could not be utilized openly to perform the evaluation process. To resolve this issue, data preparation is carried by utilizing the two methods like data wrangling and data preprocessing. Data wrangling enables the analysts to examine more complicated data more rapidly, to accomplish more precise results, and due to these, improved decisions could be made. Several companies have shifted to data wrangling due to the achievement that it has made.

Data leakage describes a mistake they are being made by the originator of a machine learning model where they mistakenly share information among the test and training datasets. Usually, when dividing a data set into testing and training sets, the aim is to make sure that no data is shared among the two. Data leakage often leads to idealistically high levels of performance on the test set, since the model is being run on data that it had previously seen—in a certain capacity—in the training set.

Data wrangling is also known as data munging, data remediation, or data cleaning, which signifies various processes planned to be converted raw information into a more easily utilized form. The particular techniques vary from project to project based on the leveraging data and the objective trying to attain.

Some illustrations of data wrangling comprise:

- Combining several data sources into one dataset for investigation
- Finding mistakes in the information (for instance, blank cells in a spreadsheet) and either deleting or filling them
- Removing data that is either irrelevant or unnecessary to the project that one is functioning with

- Detecting excessive outliers in data and either explain the inconsistencies or deleting them so that analysis can occur

Data wrangling be able to be an automatic or manual method. Scenarios in which datasets are extremely big, automatic data cleaning is becoming a must. In businesses that hire a complete data group, a data researcher or additional group representative is usually liable for data wrangling. In small businesses, nondata experts are frequently liable for cleaning their data prior to leveraging it.

5.3 Data Wrangling Stages

Individual data project demands a distinctive method to make sure their final dataset is credible and easily comprehensible ie different procedures usually notify the proposed methodology. These are often called data wrangling steps or actions shown in Figure 5.1.

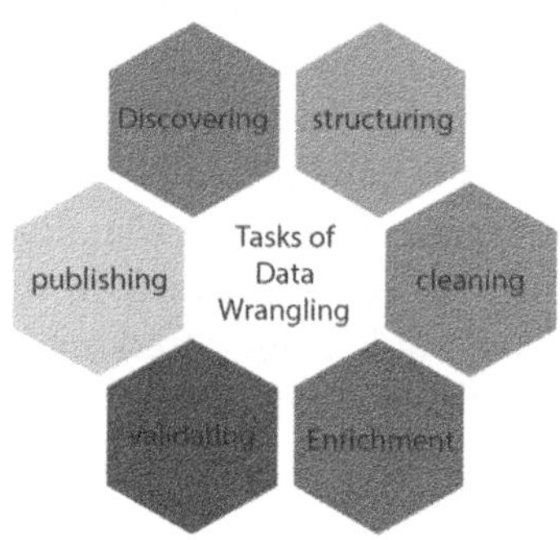

Figure 5.1 Task of data wrangling.

5.3.1 Discovery

Discovery means the method of getting acquainted with the data so one can hypothesize in what way one might utilize it. One can compare it to watching in the fridge before preparing meals to view what things are available.

During finding, one may find patterns or trends in the data, together with apparent problems, like lost or inadequate values to be resolved. This is a major phase, as it notifies each task that arises later.

5.3.2 Structuring

Raw information is usually impractical in its raw form since it is either misformatted or incomplete for its proposed application. Data structuring is the method of accepting raw data and translating it to be much more easily leveraged. The method data takes would be dependent on the analytical model that we utilize to interpret them.

5.3.3 Cleaning

Data cleaning is the method of eliminating fundamental inaccuracies in the data that can alter the review or make them less important. Cleaning is able to take place in various types, comprising the removal of empty rows or cells, regulating inputs, and eliminating outliers. The purpose of data cleaning is to make sure that there are no inaccuracies (or minimal) that might affect your last analysis.

5.3.4 Improving

When one realizes the current data and have turned it into a more useful state, one must define out if one has all of the necessary data projects at hand. If that is not the case, one might decide to enhance or strengthen the data by integrating values from additional datasets. Therefore, it is essential to realize what other information is accessible for usage.

If one determines that fortification is required, one must repeat these steps for new data.

5.3.5 Validating

Data validation is the method of checking that data is simultaneously coherent and of sufficiently high quality. Throughout the validation process, one might find problems that they want to fix or deduce that the data is ready to be examined. Generally, validation is attained by different automatic processes, and it needs to be programmed.

5.3.6 Publishing

When data is verified, one can post it. This includes creating it accessible to other people inside the organization for additional analysis. The structure one uses to distribute the data like an electronic file or written report will be based on data and organizational objectives.

5.4 Significance of Data Wrangling

Any assessments a company carries out would eventually be restricted by data notifying them. If data is inaccurate, inadequate, or incorrect, then the analysis is going to be reducing the value of any perceptions assembled.

Data wrangling aims to eliminate that possibility by making sure that the data is in a trusted state prior to it is examined and leveraged. This creates an important portion of the analytic process.

It is essential to notice that the data wrangling can be time-consuming and burdensome resources, especially once it is made physically. Therefore, several organizations establish strategies and good practices that support workers to simplify the process of data cleaning—for instance, demanding that data contain specific data or be in a certain structure formerly it has been uploaded to the database.

Therefore, it is important to realize the different phases of the data wrangling method and the adverse effects that are related to inaccurate or erroneous data.

5.5 Data Wrangling Examples

While usually performed by data researchers & technical assistants, the results of data wrangling are felt by all of us. For this part, we are concentrating on the powerful opportunities of data wrangling with Python.

For instance, data researchers will utilize data wrangling to web scraping and examine performance advertising data from a social network. This data could even be coupled with network analysis to come forward with an all-embracing matrix explaining and detecting marketing efficiency and budget costs, hence informing future pay-out distribution[14].

5.6 Data Wrangling Tools for Python

Data wrangling is the most time-consuming part of managing data and analysis for data researchers. There are multiple tools on the market to sustain the data wrangling endeavors and simplifying the process without endangering the functionality or integrity of data.

Pandas

Pandas is one of the most widely used data wrangling tools for Python. Since 2009, the open-source data analysis and manipulation tool has evolved and aims of being the "most robust and resilient open-source data analysis/manipulation tool available in every language."

Pandas' stripped-back attitude is aimed towards those with an already current level of data wrangling knowledge, as its power lies in the manual features that may not be perfect for beginners. If someone is willing to learn how to use it and to exploit its power, Pandas is the perfect solution shown in Figure 5.2.

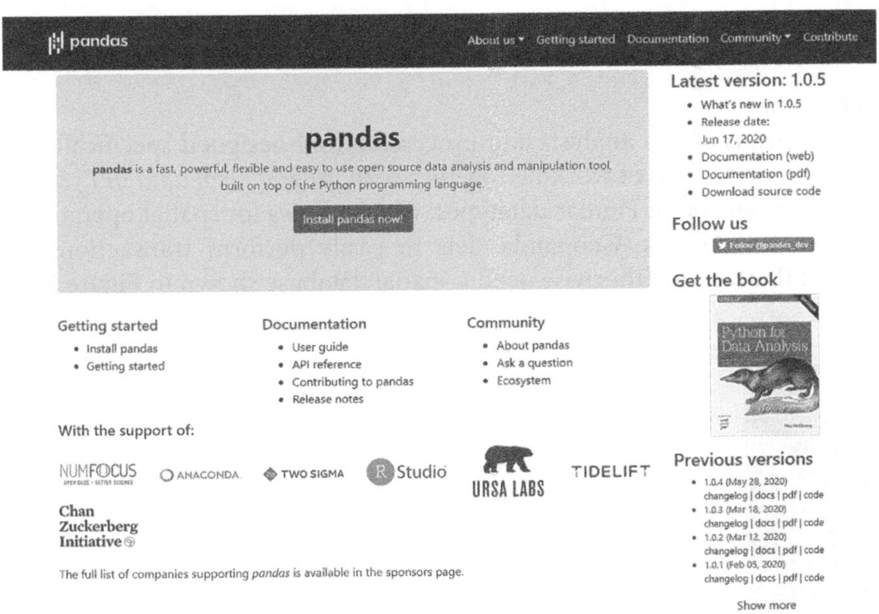

Figure 5.2 Pandas (is a software library that was written for Python programming language for data handling and analysing).

NetworkX

NetworkX is a graph data-analysis tool and is primarily utilized by data researchers. The Python package for the "setting-up, exploitation, and exploration of the structure, dynamics, and functionality of the complicated networks" can support the simplest and most complex instances and has the power to collaborate with big nonstandard datasets shown in Figure 5.3.

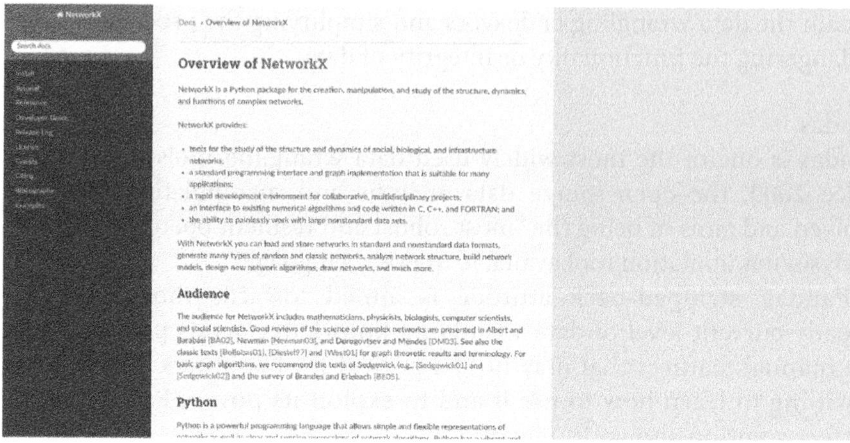

Figure 5.3 NetworkX.

Geopandas

Geopandas is a data analysis and processing tool designed specifically to simplify the process of working together with *geographical data in Python*. It is an expansion of Pandas datatypes, which allows for spatial operations on geometric kinds. Geopandas lets to easily perform transactions in Python that would otherwise need a spatial database shown in Figure 5.4.

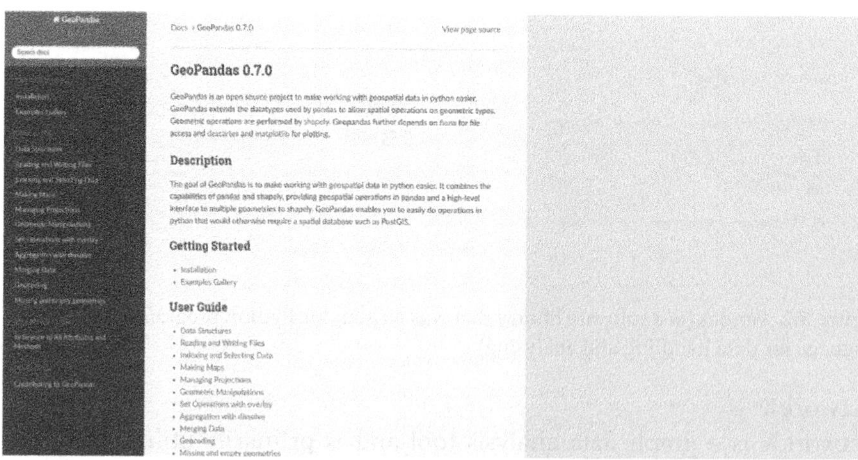

Figure 5.4 Geopandas.

Extruct
One more expert tool, Extruct is a library to extract built-in metadata from HTML markup by offering a command-line tool that allows the user to retrieve a page and extract the metadata in a quick and easy way.

5.7 Data Wrangling Tools and Methods

Multiple tools and methods can help specialists in their attempts to wrangle data so that others can utilize it to reveal insights. Some of these tools can make it easier for data processing, and others can help to make data more structured and understandable, but everyone is convenient to experts as they wrangle data to avail their organizations.

Processing and Organizing Data
A particular tool an expert uses to handle and organize information can be subject to the data type and the goal or purpose for the data. For instance, spreadsheet software or platform, like Google Sheets or Microsoft Excel, may be fit for specific data wrangling and organizing projects.

Solutions Review observes that big data processing and storage tools, like Amazon Web Services and Google BigQuery, aid in sorting and storing data. For example, Microsoft Excel can be employed to catalog data, like the number of transactions a business logged during a particular week. Though, Google BigQuery can contribute to data storage (the transactions) and can be utilized for data analysis to specify how many transactions were beyond a specific amount, periods with a specific frequency of transactions, etc.

Unsupervised and supervised machine learning algorithms can contribute to the process and examine the stored and systematized data. "In a supervised learning model, the algorithm realizes on a tagged data set, offering an answer key that the algorithm can be used to assess their accuracy on training data". "Conversely, an unsupervised model offers unlabeled data that the algorithm attempts to make any sense of by mining patterns and features on its own."

For example, an unsupervised learning algorithm could be provided 10,000 images of pizza, changing slightly in size, crust, toppings, and other factors, and attempt to make sense of those images without any existing labels or qualifiers. A supervised learning algorithm that was intended to recognize the difference between data sets of pictures of either pizza or donuts could ideally categorize through a huge data set of images of both.

Both learning algorithms would permit the data to be better organized than what was incorporated in the original set.

Cleaning and Consolidating Data

Excel permits individuals to store information. The organization Digital Vidya offers tips for cleaning data in Excel, such as removing extra spaces, converting numbers from text into numerals, and eliminating formatting. For instance, after data has been moved into an Excel spreadsheet, removing extra spaces in separate cells can help to offer more precise analytics services later on. Allowing text-written numbers to have existed (e.g., nine rather than 9) may hamper other analytical procedures.

Data wrangling best practices may vary by individual or organization who will access the data later, and the purpose or goal for the data's use. The small bakery may not have to buy a huge database server, but it might need to use a digital service or tool that is the most intuitive and inclusive than a folder on a desktop computer. Particular kinds of database systems and tools contain those offered by Oracle and MySQL.

Extracting Insights from Data

Professionals leverage various tools for extracting data insights, which take place after the wrangling process.

Descriptive, predictive, diagnostic, and prescriptive analytics can be applied to a data set that was wrangled to reveal insights. For example, descriptive analytics could reveal the small bakery how much profit is produced in a year. Descriptive analytics could explain why it generated that amount of profit. Predictive analytics could reveal that the bakery may also see a 10% decrease in profit over the coming year. Prescriptive analytics could emphasize potential solutions that may help the bakery alleviate the potential drop.

Datamation also notes various kinds of data tools that can be beneficial to organizations. For example, Tableau allows users to access visualizations of their data, and IBM Cognos Analytics offers services that can help in different stages of an analytics process.

5.8 Use of Data Preprocessing

Data preprocessing is needed due to the existence of unformatted real-world data. Predominantly real-world data is made up of

Missing data (Inaccurate data) —There are several causes for missing data like data is not continually gathered, an error in data entry, specialized issues with biometric information, and so on.

The existence of noisy data (outliers and incorrect data)—The causes for the presence of noisy data might be a technical challenge of tools that collect data, a human error when entering data, and more.

Data Inconsistency—The presence of data inconsistency is because of the presence of replication within data, dataentry, that contains errors in names or codes i.e., violation of data restrictions, and so on. In order to process raw data, data preprocessing is carried out shown in Figure 5.5.

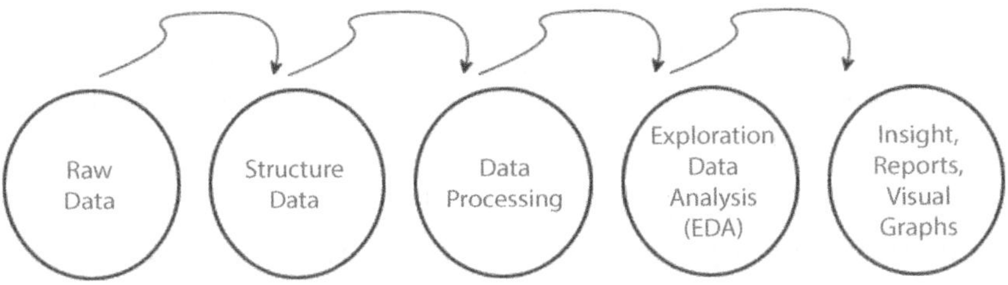

Figure 5.5 Data processing in Python.

5.9 Use of Data Wrangling

While implementing deep learning and machine learning, data wrangling is utilized to manipulate the problem of data leakage.

Data leakage in deep learning/machine learning

Because of the overoptimization of the applied model, data leakage leads to an invalid deep learning/machine learning model. Data leakage is a term utilized once the data from the exterior, i.e., not portion of the training dataset is utilized for the learning method of the model. This extra learning of data by the applied model will negate the calculated estimated efficiency of the model [9].

For instance, once we need to utilize the specific feature to perform Predictive Analysis, but that particular aspect does not exist at the moment of training dataset then data leakage would be created within the model. Leakage of data could be shown in several methods that are listed below:

- Data Leakage for the training dataset from the test dataset.
- Leakage of the calculated right calculation to the training dataset.

- Leakage of upcoming data into the historical data.
- Utilization of data besides the extent of the applied algorithm.

The data leakage has been noted from the two major causes of deep learning/machine learning algorithms like training datasets and feature attributes (variables) [10].

Leakage of data is noted at the moment of the utilization of complex datasets. They are discussed later:

- The dataset is a difficult problem while splitting the time series dataset into test and training.
- Enactment of sampling in a graphic issue is a complicated task.
- Analog observations storage is in the type of images and audios in different files that have a specified timestamp and size.

Performance of data preprocessing
Data pretreatment is performed to delete the reason of raw real-world data and lost data to handle [11]. Following three distinct steps can be performed,

- Ignores the Inaccurate record—It is the most simple and effective technique to manage inaccurate data. But this technique must not be carried out once the number of inaccurate data is massive or if the pattern of data is associated with an unidentified fundamental root of the cause of the statement problem.
- Filling the lost value by hand—It is one of the most excellent-selected techniques. But there is one constraint that once there is a big dataset and inaccurate values are big after that, this methodology is effective as it will be a time-consuming task.
- Filling utilizing a calculated value—The inaccurate values can be filled out by calculating the median, mean, or mode of the noted certain values. The different methods might be the analytical values that are calculated by utilizing any algorithm of deep learning or machine learning. But one disadvantage of this methodology is that it is able to produce systematic errors within the data as computed values are inaccurate regarding the noted values.

Process of handling the noisy data.
A method that can be followed are specified below:

- Machine learning—This can be performed on the data smoothing. For instance, a regression algorithm is able to be utilized to smooth data utilizing a particular linear function.
- Clustering method—In this method, the outliers can be identified by categorizing the related information in a similar class, i.e., in a similar cluster.
- Binning method—In this technique, data sorting is achieved regarding the desired values of the vicinity. This technique is also called local smoothing.
- Removing manually—The noisy data can be taken off by hand by humans, but it is a time-consuming method so largely this approach is not granted precedence.
- The contradictory data is managed to utilize the external links and knowledge design tools such as the knowledge engineering process.

Data Leakage in Machine Learning
The leakage of data can make to generate overly enthusiastic if not entirely invalid prediction models. The leakage of data is as soon as information obtained externally from the training dataset is utilized to build the model [12]. This extra information may permit the model to know or learn anything that it otherwise would not know and in turn, invalidating the assessed efficiency of the model which is being built.

This is a major issue for at least three purposes:

1. It is a challenge if one runs a machine learning contest. The leaky data is applied in the best models instead of being a fine generic model of the basic issue.
2. It is an issue while one is a company that provides your data. Changing an obfuscation and anonymization can lead to a privacy breach that you never expected.
3. It is an issue when one develops their own forecasting model. One might be making overly enthusiastic models, which are sensibly worthless and may not be utilized in manufacturing.

To defeat there are two fine methods that you can utilize to reduce data leakage while evolving predictive models are as follows:

1. Carry out preparation of data within the cross-validation folds.
2. Withhold a validation dataset for final sanity checks of established models.

Performing Data Preparation Within Cross-Validation Folds

While data preparation of data, leakage of information in machine learning may also take place. The impact is overfitting the training data, and which has an overly enthusiastic assessment of the model's efficiency on undetected data. To standardize or normalize the whole dataset, one could sin leakage of data then cross-validation has been utilized to assess the efficiency of the model.

The method of rescaling data that one carried out had an understanding of the entire distribution of data in the training dataset while computing the scaling parameters (such as mean and standard deviation or max and min). This knowledge was postmarked rescaled values and operated by all algorithms in a cross-validation test harness [13].

In this case, a nonleaking assessment of machine learning algorithms would compute the factors for data rescaling within every folding of the cross-validation and utilize these factors to formulate the data on every cycle on the held-out test fold. To recompute or reprepare any necessary data preparation within cross-validation folds comprising tasks such as removal or outlier, encoding, selection of feature, scaling feature and projection techniques for dimensional reduction, and so on.

Hold Back a Validation Dataset

An easier way is to divide the dataset of training into train and authenticate the sets and keep away the validation dataset. After the completion of modeling processes and actually made-up final model, assess it on the validation dataset. This might provide a sanity check to find out if the estimation of performance is too enthusiastic and was disclosed.

5.10 Data Wrangling in Machine Learning

The establishment of automatic solutions for data wrangling deals with one most important hurdle: the cleaning of data needs intelligence and not a simple reiteration of work. Data wrangling is meant by having a grasp of exactly what does the user seeks to solve the differences between data sources or say, the transformation of units.

A standard wrangling operation includes these steps: mining of the raw information from sources, the usage of an algorithm to explain the raw data into predefined data structures, and transferring the findings into a data mart for storing and upcoming use.

At present, one of the greatest challenges in machine learning remains in computerizing data wrangling. One of the most important obstacles is data leakage, i.e., throughout the training of the predictive model utilizing ML, it utilizes data outside of the training data set, which is not verified and unmarked.

The few data-wrangling automation software currently available utilize peer-to-peer ML pipelines. But those are far away and a few in-between. The market definitely needs additional automated data wrangling programs.

These are various types of machine learning algorithms:

- Supervised ML: utilized to standardize and consolidate separate data sources.
- Classification: utilized in order to detect familiar patterns.
- Normalization: utilized to reorganize data into the appropriate manner.
- Unsupervised ML: utilized for research of unmarked data

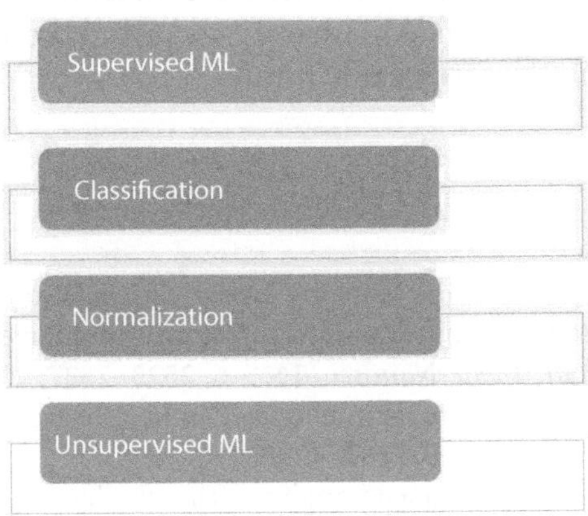

Figure 5.6 Various types of machine learning algorithms.

As it is, a large majority of businesses are still in the initial phases of the implementation of AI for data analytics. They are faced with multiple obstacles: expenses, tackling data in silos, and the fact that it really is not simple for business analysts—those who do not need an engineering or data science experience—to better understand machine learning shown in Figure 5.6.

5.11 Enhancement of Express Analytics Using Data Wrangling Process

Our many years of experience in dealing with data demonstrated that the data wrangling process is the most significant initial step in data analytics. Our data wrangling process involves all the six tasks like data discovery, (listed above), etc, in order to formulate the enterprise data for the analysis. The data wrangling process will help to discover intelligence within the most different data sources. We will correct human mistakes in collecting and tagging data and also authenticate every data source.

5.12 Conclusion

Finding data leakage in advance and revising for it is a vital part of an improvement in the definition of a machine learning issue. Multiple types of leakage are delicate and are best perceived by attempting to extract features and train modern algorithms on the problem. Data wrangling and data leakage are being handled to identify and avoid the additional process in health services in the foreseeable future.

References

1. Basheer, S. *et al.*, Machine learning based classification of cervical cancer using K-nearest neighbour, random forest and multilayer perceptron algorithms. *J. Comput. Theor. Nanosci.*, 16, 5-6, 2523–2527, 2019.
2. Deekshaa, K., Use of artificial intelligence in healthcare and medicine, *Int. J. Innov. Eng. Res. Technol.*, 5, 12, 1–4. 2021.
3. Terrizzano, I.G. *et al.*, Data wrangling: The challenging journey from the wild to the lake. *CIDR*, 2015.
4. Joseph, M. Hellerstein, T. R., Heer, J., Kandel, S., Carreras, C., Principles of data wrangling, Publisher(s): O'Reilly Media, Inc. ISBN: 9781491938928 July 2017.
5. Quinto, B., Big data visualization and data wrangling, in: *Next-Generation Big Data*, pp. 407–476, Apress, Berkeley, CA, 2018.

6. McKinney, W., Python for data analysis, Publisher(s): O'Reilly Media, Inc. ISBN: 9781491957660 October 2017.
7. Koehler, M. *et al.*, Data context informed data wrangling. *2017 IEEE International Conference on Big Data (Big Data)*, IEEE, 2017.
8. Kazil, J. and Jarmul, K., Data wrangling with Python Publisher(s): O'Reilly Media, Inc. ISBN: 9781491948774 February 2016
9. Sampaio, S. *et al.*, A conceptual approach for supporting traffic data wrangling tasks. *Comput. J.*, 62, 3, 461–480, 2019.
10. Jiang, S. and Kahn, J., Data wrangling practices and collaborative interactions with aggregated data. *Int. J. Comput.-Support. Collab. Learn.*, 15, 3, 257–281, 2020.
11. Azeroual, O., Data wrangling in database systems: Purging of dirty data. *Data*, 5, 2, 50, 2020.
12. Patil, M.M. and Hiremath, B.N., A systematic study of data wrangling. *Int. J. Inf. Technol. Comput. Sci.*, 1, 32–39, 2018.
13. Konstantinou, N. *et al.*, The VADA architecture for cost-effective data wrangling. *Proceedings of the 2017 ACM International Conference on Management of Data*, 2017.
14. Swetha, K.R., Niranjanamurthy, M., Amulya, M.P., Manu, Y.M., Prediction of pneumonia using big data, deep learning and machine learning techniques. *2021 6th International Conference on Communication and Electronics Systems (ICCES)*, pp. 1697–1700, 2021, doi: 10.1109/ICCES51350.2021.9489188.

6

Importance of Data Wrangling in Industry 4.0

Rachna Jain[1], Geetika Dhand[2], Kavita Sheoran[2] and Nisha Aggarwal[2*]

[1]JSS Academy of Technical Education, Noida, India
[2]Maharaja Surajmal Institute of Technology, New Delhi, India

Abstract
There is tremendous growth in data in this industry 4.0 because of vast amount of information. This messy data need to be cleaned in order to provide meaningful information. Data wrangling is a method of converting this messy data into some useful form. The main aim of this process is to make stronger intelligence after collecting input from many sources. It helps in providing accurate data analysis, which leads to correct decisions in developing businesses. It even reduces time spent, which is wasted in analysis of haphazard data. Better decision skills are driven from management due to organized data. Key steps in data wrangling are collection or acquisition of data, combining data for further use and data cleaning which involves removal of wrong data. Spreadsheets are powerful method but not making today's requirements. Data wrangling helps in obtaining, manipulating and analyzing data. R language helps in data management using different packages dplyr, httr, tidyr, and readr. Python includes different data handling libraries such as numpy, Pandas, Matplotlib, Plotly, and Theano. Important tasks to be performed by various data wrangling techniques are cleaning and structuring of data, enrichment, discovering, validating data, and finally publishing of data.

Data wrangling includes many requirements like basic size encoding format of the data, quality of data, linking and merging of data to provide meaningful information. Major data analysis techniques include data mining, which extracts information using key words and patterns, statistical techniques include computation of mean, median, etc. to provide an insight into the data. Diagnostic analysis involves pattern recognition techniques to answer meaningful questions, whereas predictive analysis includes forecasting the situations so that answers help in yielding meaningful strategies for an organization. Different data wrangling tools include

**Corresponding author*: nishaa@mait.ac.in

M. Niranjanamurthy, Kavita Sheoran, Geetika Dhand, and Prabhjot Kaur (eds.) Data Wrangling: Concepts, Applications and Tools, (109–122) © 2023 Scrivener Publishing LLC

excel query/spreadsheet, open refine having feature procurement, Google data prep for exploration of data, tabula for all kind of data applications and CSVkit for converting data. Thus, data analysis provides crucial decisions for an organization or industry. It has its applications in vast range of industries including healthcare and retail industries. In this chapter, we will summarize major data wrangling techniques along with its applications in different areas across the domains.

Keywords: Data wrangling, data analysis, industry 4.0, data applications, Google Sheets, industry

6.1 Introduction

Data Deluge is the term used for explosion of data. Meaningful information can be extracted from raw data by conceptualizing and analyzing data properly. Data Lake is the meaningful centralized repository made from raw data to do analytical activities [1]. In today's world, every device that is connected to the internet generates enormous amount of data. A connected plane generates 5 Tera Byte of data per day, connected car generates 4 TB of data per day. A connected factory generates 5 Penta Byte of data per day. This data has to be organized properly to retrieve meaningful information from it. Data management refers to data modeling and management of metadata.

Data wrangling is the act of cleaning, organizing, and enriching raw data so that it can be utilized for decision making rapidly. Raw data refers to information in a repository that has not yet been processed or incorporated into a system. It can take the shape of text, graphics, or database records, among other things. The most time-consuming part of data processing is data wrangling, often known as data munging. According to data analysts, it can take up to 75% of their time to complete. It is time-consuming since accuracy is critical because this data is gathered from a variety of sources and then used by automation tools for machine learning.

6.1.1 Data Wrangling Entails

a) Bringing data from several sources together in one place
b) Putting the data together
c) Cleaning the data to account for missing components or errors

Data wrangling refers to iterative exploration of data, which further refers to analysis [2]. Integration and cleaning of data has been the issue in research community from long time [3]. Basic features of any dataset are that while approaching dataset for the first-time size and encoding has to be explored. Data Quality is the central aspect of data projects. Data quality has to be maintained while documenting the data. Merging & Linking of data is another important tasks in data management. Documentation & Reproducibility of data is also equally important in the industry [4].

Data wrangling is essential in the most fundamental sense since it is the only method to convert raw data into useful information. In a practical business environment, customer or financial information typically comes in pieces from different departments. This data is sometimes kept on many computers, in multiple spreadsheets, and on various systems, including legacy systems, resulting in data duplication, erroneous data, or data that cannot be found to be utilized. It is preferable to have all data in one place so that you can get a full picture of what is going on in your organization [5].

6.2 Steps in Data Wrangling

While data wrangling is the most critical initial stage in data analysis, it is also the most tiresome, it is frequently stated that it is the most overlooked. There are six main procedures to follow when preparing data for analysis as part of data munging [6].

- Data Discovery: This is a broad word that refers to figuring out what your data is all about. You familiarize yourself with your data in this initial stage.
- Data Organization: When you first collect raw data, it comes in all shapes and sizes, with no discernible pattern. This data must be reformatted to fit the analytical model that your company intends to use [7].
- Data Cleaning: Raw data contains inaccuracies that must be corrected before moving on to the next stage. Cleaning entails addressing outliers, making changes, or altogether erasing bad data [8].
- Data Enrichment: At this point, you have probably gotten to know the data you are working with. Now is the moment to consider whether or not you need to embellish the basic data [9].
- Data Validation: This activity identifies data quality problems, which must be resolved with the appropriate transformations [10]. Validation rules necessitate repetitive programming procedures to ensure the integrity and quality of your data.
- Data Publishing: After completing all of the preceding processes, the final product of your data wrangling efforts is pushed downstream for your analytics requirements.

Data wrangling is an iterative process that generates the cleanest, most valuable data before you begin your analysis [11]. Figure 6.1 displays that how messy data can be converted into useful information.

This is an iterative procedure that should result in a clean and useful data set that can then be analyzed [12]. This is a time-consuming yet beneficial technique since it helps analysts to extract information from a big quantity of data that would otherwise be unreadable. Figure 6.2 shows the organized data using data wrangling.

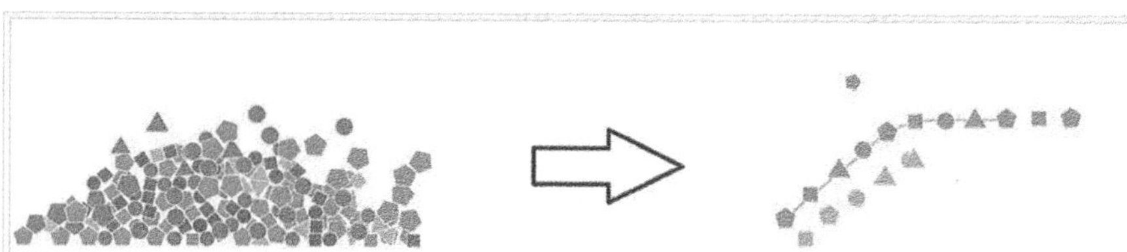

Figure 6.1 Turning messy data into useful statistics.

Starting Data

Name	Phone	Birth Date	State
John, Smith	445-881-4478	August 12, 1989	Maine
Jennifer Tal	+1-189-456-4513	11/12/1965	Tx
Gates, Bill	(876)546-8165	June 15, 72	Kansas
Alan Fitch	5493156648	2-6-1985	Oh
Jacob Alan	156-4896	January 3	Alabama

Result

Name	Phone	Birth Date	State
John Smith	445-881-4478	1989-08-12	Maine
Jennifer Tal	189-456-4513	1965-11-12	Texas
Bill Gates	876-546-8165	1972-06-15	Kansas
Alan Fitch	549-315-6648	1985-02-06	Ohio

Figure 6.2 Organized data using data wrangling.

6.2.1 Obstacles Surrounding Data Wrangling

In contrast to data analytics, about 80% of effort is lost in gaining value from big data through data wrangling [13]. As a result, efficiency must improve. Until now, the challenges of big data with data wrangling have been solved on a phased basis, such as data extraction and integration. Continuing to disseminate knowledge in the areas with the greatest potential to improve the data wrangling process. These challenges can only be met on an individual basis.

- Any data scientist or data analyst can benefit from having direct access to the data they need. Otherwise, we must provide brief orders in order to obtain "scrubbed" data, with the goal of granting the request and ensuring appropriate execution [14]. It is difficult and time-consuming to navigate through the policy maze.
- Machine Learning suffers from data leaking, which is a huge problem to solve. As Machine Learning algorithms are used in data processing, the risks increase gradually. Data accuracy is a crucial component of prediction [15].
- Recognizing the requirement to scale queries that can be accessed with correct indexing poses a problem. Before constructing a model, it is critical to thoroughly examine the correlation. Before assessing the relationship to the final outcome, redundant and superfluous data must be deleted [16]. Avoiding this would be fatal in the long run. Frequently, in huge data sets of files, a cluster of closely related columns appears, indicating that the data is redundant and making model selection more difficult. Despite the fact that these repeatednesses will offer a significant correlation coefficient, it will not always do so [17].
- There are a few main difficulties that must be addressed. For example, different quality evaluations are not limited, and even simple searches used in mappings would necessitate huge updates to standard expectations in the case of a large dataset [18]. A dataset is frequently devoid of values, has errors, and contains noise. Some of the causes include soapy eye, inadvertent mislabeling, and technical flaws. It has a well-known impact on the class of data processing tasks, resulting in subpar outputs and, ultimately, poorly managed business activity [19]. In ML algorithms, messy, unrealistic

data is like rubbing salt in the wounds. It is possible that a trained dataset algorithm will be unsuitable for its purposes.

- Reproducibility and documentation are critical components of any study, but they are frequently overlooked [20]. Data processing and procedures across time, as well as the regeneration of previously acquired conclusions, are mutual requirements that are challenging to meet, particularly in mutually interacting connectivity [21].
- Selection bias is not given the attention it deserves until a model fails. It is very important in data science. It is critical to make sure the training data model is representative of the operating model [22]. In bootstrapped design, ensuring adequate weights necessitates building a design specifically for this use.
- Data combining and data integration are frequently required to construct the image. As a result, merging, linking divergent designs, coding procedures, rules, and modeling data are critical as we prepare data for later use [23].

6.3 Data Wrangling Goals

1. Reduce Time: Data analysts spend a large portion of their time wrangling data, as previously indicated. It consumes much of the time of some people. Consider putting together data from several sources and manually filling in the gaps [24]. Alternatively, even if code is used, stringing it together accurately takes a long time. Solvexia, for example, can automate 10× productivity.
2. Data analysts can focus on analysis: Once a data analyst has freed up all of the time they would have spent wrangling data, they can use the data to focus on why they were employed in the first place—to perform analysis [25]. Data analytics and reporting may be produced in a matter of seconds using automation techniques.
3. Decision making that is more accurate and takes less time: Information must be available quickly to make business decisions [26]. You can quickly make the best decision possible by utilizing automated technologies for data wrangling and analytics.

4. More in-depth intelligence: Data is used in every facet of business, and it will have an impact on every department, from sales to marketing to finance [27]. You will be able to better comprehend the present state of your organization by utilizing data and data wrangling, and you will be able to concentrate your efforts on the areas where problems exist.
5. Data that is accurate and actionable: You will have ease of mind knowing that your data is accurate, and you will be able to rely on it to take action, thanks to proper data wrangling [28].

6.4 Tools and Techniques of Data Wrangling

It has been discovered that roughly 80% of data analysts spend the majority of their time wrangling data rather than doing actual analysis. Data wranglers are frequently employed if they possess one or more of the following abilities: Knowledge of a statistical language, such as R or Python, as well as SQL, PHP, Scala, and other programming languages.

6.4.1 Basic Data Munging Tools

- Excel Power Query/Spreadsheets — the most basic structuring tool for manual wrangling.
- OpenRefine — more sophisticated solutions, requires programming skills
- Google DatePrep — for exploration, cleaning, and preparation.
- Tabula — swiss army knife solutions — suitable for all types of data
- DataWrangler — for data cleaning and transformation.
- CSVKit — for data converting

6.4.2 Data Wrangling in Python

1. Numpy (aka Numerical Python) — The most basic package is Numpy (also known as Numerical Python). Python has a lot of capabilities for working with n-arrays and matrices. On the NumPy array type, the library enables vectorization of mathematical operations, which increases efficiency and speeds up execution.

2. Pandas — intended for quick and simple data analysis. This is particularly useful for data structures with labelled axes. Explicit data alignment eliminates typical mistakes caused by mismatched data from many sources.
3. Matplotlib — Matplotlib is a visualisation package for Python. Line graphs, pie charts, histograms, and other professional-grade figures benefit from this.
4. Plotly — for interactive graphs of publishing quality. Line plots, scatter plots, area charts, bar charts, error bars, box plots, histograms, heatmaps, subplots, multiple-axis, polar graphs, and bubble charts are all examples of useful graphics.
5. Theano — Theano is a numerical computing library comparable to Numpy. This library is intended for quickly defining, optimising, and evaluating multi-dimensional array mathematical expressions.

6.4.3 Data Wrangling in R

1. Dplyr — a must-have R tool for data munging. The best tool for data framing. This is very handy when working with data in categories.
2. Purrr — useful for error-checking and list function operations.
3. Splitstackshape — a tried-and-true classic. It is useful for simplifying the display of complicated data sets.
4. JSOnline — a user-friendly parsing tool.
5. Magrittr — useful for managing disjointed sets and putting them together in a more logical manner.

6.5 Ways for Effective Data Wrangling

Data integration, based on current ideas and a transitional data cleansing technique, has the ability to improve wrapped inductive value.

Manually wrangling data or data munging allows us to manually open, inspect, cleanse, manipulate, test, and distribute data. It would first provide a lot of quick and unreliable data [29]. However, because of its inefficiency, this practice is not recommended. In single-case current analysis instances, this technique is critical. Long-term continuation of this procedure takes a lot of time and is prone to error owing to human participation. This method always has the risk of overlooking a critical phase, resulting in inaccurate data for the consumers [30].

To make matter better, we now have program-based devices that have the ability to improve data wrangling. SQL is an excellent example of a semiautomated method [31]. When opposed to a spreadsheet, one must extract data from the source into a table, which puts one in a better position for data profiling, evaluating inclinations, altering data, and executing data and presenting summary from queries within it [32]. Also, if you have a repeating command with a limited number of data origins, you can use SQL to design a process for evaluating your data wrangling [33]. Further advancement, ETL tools are a step forward in comparison to stored procedures [34]. ETLs extract data from a source form, alter it to match the consequent format, and then load it into the resultant area. Extraction-transformation-load possesses a diverse set of tools. Only a few of them are free. When compared to Standard Query Language stored queries, these tools provide an upgrade because the data handling is more efficient and simply superior. In composite transformations and lookups, ETLs are more efficient. They also offer stronger memory management capabilities, which are critical in large datasets [35].

When there is a need for duplicate and compound data wrangling, constructing a company warehouse of data with the help of completely automated workflows should be seriously considered. The technique that follows combines data wrangling with a reusable and automated mentality. This method then executes in an automated plan for current data load from a current data source in an appropriate format. Despite the fact that this method involves more thorough analysis, framework, and adjustments, as well as current data maintenance and governance, it offers the benefits of reusing extraction-transformation-load logic, and we may rework the adapted data in a number of business scenarios [36].

Data manipulation is critical in any firm research and should not be overlooked. Building timetable automated based chores to get the most out of data wrangling, adapting various data parts in a similar format saving the analysts time to deliver enhanced data combined commands is an ideal scenario for managing ones disruptive data.

6.5.1 Ways to Enhance Data Wrangling Pace

- These solutions are promising, but we must concentrate on accelerating the critical data wrangling process. It cannot be afforded to lose speed in data manipulation, so necessary measures must be taken to improve performance.
- It is difficult to emphasize the needs to the important concerns to be handled at any given time. It would also be

necessary to get quick results. The best way to cope with these problems will be described later. Each problem must be isolated in order to discover the best answer. There is a need to create some high-value factors and treat them with greater urgency. We must keep track of duties and solutions in order to speed up the process of developing a solid strategy.

- Assimilation of data specialists from industries other than the IT sector exemplifies a trend that today's businesses are not encouraging, which has resulted in a trend that modern-day firms have abandoned, resulting in the issues that have arisen. Even while data thrives for analysis, it is reliant on the function of an expert by modelling our data, which is different from data about data.
- There must be an incentive to be part of a connected society and to examine diverse case studies in your sector. Analyzing the performance of your coworkers is an excellent example of how to improve.
- Joining communities that care about each other could help you learn faster. We gain a lot of familiarity with a community of people that are determined to advance their careers in data science by constantly learning and developing on a daily basis. With the passage of time, we have gained more knowledge through evaluating many examples. They have the potential to be extremely important.
- Every crew in a corporation has its own goals and objectives. However, they all have the same purpose in mind. Collaboration with other teams, whether engineering, data science, or various divisions within a team, can be undervalued but crucial. It brings with it a new way of thinking. We are often stuck in a rut, and all we need is a slight shift in viewpoint. For example, the demand to comprehend user difficulties may belong in the gadget development team, rather than in the thoughts of the operations team, because it might reduce the amount of time spent on logistics. As a result, collaboration could speed up the process of locating the perfect dataset.
- Data errors are a well-known cause of delays, and they are caused by data mapping, which is extremely challenging in the case of data wrangling. Data manipulation is one answer to this problem. It does not appear to be a realistic solution,

but it does lessen the amount of time we spend mapping our data. Data laboratories are critical in situations when an analyst has the opportunity to use potential data streams, as well as variables to determine whether they are projecting or essential in evaluating or modeling the data.

- When data wrangling is used to gather user perceptions with the help of Face book, Twitter, or any other social media, polls, and comment sections, it enhances knowledge of how to use data appropriately, such as user retention. However, the complexity increases when the data wrangle usage is not identified. The final outcome obtained through data wrangling would be unsatisfactory. As a result, it is critical to extract the final goal via data wrangling while also speeding up the process.
- Intelligent awareness has the ability to extract information and propose solutions to data wrangling issues. We must determine whether scalability and granularity are maintained and respond appropriately. Try to come up with a solution for combining similar datasets throughout different time periods. Find the right gadgets or tools to help you save time when it comes to data wrangling. We need to know if we can put in the right structure with the least amount of adjustments. To improve data wrangling, we must examine findings.
- The ability to locate key data in order to make critical decisions at the correct time is critical in every industry. Randomness or complacency has no place in a successful firm, and absolute data conciseness is required.

6.6 Future Directions

Quality of data, merging of different sources is the first phase of data handling. Heterogeneity of data is the problem faced by different departments in an organization. Data might be collected from outside sources. Analyzing data collected from different sources could be a difficult task. Quality of data has to be managed properly since different organization yield content rich in information but quality of data becomes poor. This research paper gave a brief idea about toolbox from the perspective of a data scientist that will help in retrieving meaningful information. Brief overview of tools related to data wrangling has been covered in the paper.

Practical applications of R language, RStudio, Github, Python, and basic data handling tools have been thoroughly analyzed. User can implement statistical computing by reading data either in CSV kit or in python library and can analyze data using different functions. Exploratory data analysis techniques are also important in visualizing data graphics. This chapter provides a brief overview of different toolset available with a data scientist. Further, it can be extended for data wrangling using artificial intelligence methods.

References

1. Terrizzano, I.G., Schwarz, P.M., Roth, M., Colino, J.E., Data wrangling: The challenging yourney from the wild to the lake, in: *CIDR*, January 2015.
2. Furche, T., Gottlob, G., Libkin, L., Orsi, G., Paton, N.W., Data wrangling for big data: Challenges and opportunities, in: *EDBT*, pp. 473–478, March 2016.
3. Kandel, S., Heer, J., Plaisant, C., Kennedy, J., Van Ham, F., Riche, N.H., Weaver, C., Lee, B., Brodbeck, D., Buono, P., Research directions in data wrangling: Visualizations and transformations for usable and credible data. *Inf. Vis.*, 10, 4, 271–288, 2011.
4. Endel, F. and Piringer, H., Data wrangling: Making data useful again. *IFAC-PapersOnLine*, 48, 1, 111–112, 2015.
5. Dasu, T. and Johnson, T., *Exploratory Data Mining and Data Cleaning*, vol. 479, John Wiley & Sons, 2003.
6. https://www.bernardmarr.com/default.asp?contentID=1442 [Date: 11/11/2021]
7. Freeland, S.L. and Handy, B.N., Data analysis with the solarsoft system. *Sol. Phys.*, 182, 2, 497–500, 1998.
8. Brandt, S. and Brandt, S., *Data Analysis*, Springer-Verlag, 1998.
9. Berthold, M. and Hand, D.J., *Intelligent Data Analysis*, vol. 2, Springer, Berlin, 2003.
10. Tukey, J.W., The future of data analysis. *Ann. Math. Stat*, 33, 1, 1–67, 1962.
11. Rice, J.A., *Mathematical Statistics and Data Analysis*, Cengage Learning, 2006.
12. Fruscione, A., McDowell, J.C., Allen, G.E., Brickhouse, N.S., Burke, D.J., Davis, J.E., Wise, M., CIAO: Chandra's data analysis system, in: *Observatory Operations: Strategies, Processes, and Systems*, vol. 6270p, International Society for Optics and Photonics, June 2006.
13. Heeringa, S.G., West, B.T., Berglund, P.A., *Applied Survey Data Analysis*, Chapman and Hall/CRC, New York, 2017.
14. Carpineto, C. and Romano, G., *Concept Data Analysis: Theory and Applications*, John Wiley & Sons, 2004.
15. Swan, A.R. and Sandilands, M., Introduction to geological data analysis. *Int. J. Rock Mech. Min. Sci. Geomech. Abstr.*, 8, 32, 387A, 1995.

16. Cowan, G., *Statistical Data Analysis*, Oxford University Press, 1998.
17. Bryman, A. and Hardy, M.A. (eds.), *Handbook of Data Analysis*, Sage, 2004.
18. Bendat, J.S. and Piersol, A.G., *Random Data: Analysis and Measurement Procedures*, vol. 729, John Wiley & Sons, 2011.
19. Ott, R.L. and Longnecker, M.T., *An Introduction to Statistical Methods and Data Analysis*, Cengage Learning, 2015.
20. Nelson, W.B., *Applied Life Data Analysis*, vol. 521, John Wiley & Sons, 2003.
21. Hair, J.F. *et al.*, *Multivariate Data Analysis: A global perspective, 7th ed.*, Upper Saddle River, Prentice Hall, 2009.
22. Gelman, A., Carlin, J.B., Stern, H.S., Rubin, D.B., *Bayesian Data Analysis*, Chapman and Hall/CRC, New York, 1995.
23. Rabiee, F., Focus-group interview and data analysis. *Proc. Nutr. Soc.*, 63, 4, 655–660, 2004.
24. Agresti, A., Categorical data analysis, vol. 482, John Wiley & Sons, 2003.
25. Davis, J.C. and Sampson, R.J., *Statistics and Data Analysis in Geology*, vol. 646, Wiley, New York, 1986.
26. Van de Vijver, F. and Leung, K., Methods and data analysis of comparative research, Allyn & Bacon, 1997.
27. Daley, R., *Atmospheric Data Analysis*, Cambridge University Press, 1993.
28. Bolger, N., Kenny, D.A., Kashy, D., Data analysis in social psychology, in: *Handbook of Social Psychology*, vol. 1, pp. 233–65, 1998.
29. Bailey, T.C. and Gatrell, A.C., *Interactive Spatial Data Analysis*, vol. 413, Longman Scientific & Technical, Essex, 1995.
30. Pavlo, A., Paulson, E., Rasin, A., Abadi, D.J., DeWitt, D.J., Madden, S., Stonebraker, M., A comparison of approaches to large-scale data analysis, in: *Proceedings of the 2009 ACM SIGMOD International Conference on Management of Data*, pp. 165–178, June 2009.
31. Eriksson, L., Byrne, T., Johansson, E., Trygg, J., Vikström, C., *Multi-and Megavariate Data Analysis Basic Principles and Applications*, vol. 1, Umetrics Academy, 2013.
32. Eriksson, L., Byrne, T., Johansson, E., Trygg, J., Vikström, C., *Multi-and Megavariate Data Analysis Basic Principles and Applications*, vol. 1, Umetrics Academy, 2013.
33. Hedeker, D. and Gibbons, R.D., Longitudinal data analysis, Wiley-Interscience, 2006.
34. Ilijason, R., ETL and advanced data wrangling, in: *Beginning Apache Spark Using Azure Databricks*, pp. 139–175, Apress, Berkeley, CA, 2020.
35. Rattenbury, T., Hellerstein, J.M., Heer, J., Kandel, S., Carreras, C., *Principles of Data Wrangling: Practical Techniques for Data Preparation*, O'Reilly Media, Inc, 2017.
36. Koehler, M., Abel, E., Bogatu, A., Civili, C., Mazilu, L., Konstantinou, N., ... Paton, N.W., Incorporating data context to cost-effectively automate end-to-end data wrangling. *IEEE Trans. Big Data*, 7, 1, 169–186, 2019.

7

Managing Data Structure in R

Mittal Desai[1]* and Chetan Dudhagara[2]

[1]Smt. Chandaben Mohanbhai Patel Institute of Computer Applications, Charotar University of Science and Technology, Changa, Anand, Gujarat, India
[2]Dept. of Communication & Information Technology, International Agribusiness Management Institute, Anand Agricultural University, Anand, Gujarat, India

Abstract
The data structure allowed us to organize and store the data in a way that we needed in our applications. It helps us to reduce the storage space in a memory and fast access of data for various tasks or operations. R provides an interactive environment for data analysis and statistical computing. It supports several basic various data types that are frequently used in different calculation and analysis-related work. It supports six basic data types, such as numeric (real or decimal), integer, character, logical, complex, and raw. These basic data types are used for its analytics-related works on data. There are few more efficient data structures available in R, such as Vector, Factor, Matrix, Array, List, and Dataframe.

Keywords: Data structure, vector, factor, array, list, data frame

7.1 Introduction to Data Structure

R is an open-source programming language and software environment that is widely used as a statistical software and data analysis tool. R provides a wide variety of statistical and graphical techniques, including linear and nonlinear modeling, statistical tests, time-series analysis, classification, clustering, etc. [3].

The data structure is a way of organizing and storing the data in a memory device so that it can be used efficiently to perform various tasks on it.

**Corresponding author*: bhattmittal2008@gmail.com

M. Niranjanamurthy, Kavita Sheoran, Geetika Dhand, and Prabhjot Kaur (eds.) Data Wrangling: Concepts, Applications and Tools, (123–146) © 2023 Scrivener Publishing LLC

R supports several basic data types that are frequently used in different calculation. It has six primitive data types, such as numeric (real or decimal), integer, character, logical, complex, and raw [4].

The data structure is often organized by their dimensionality, such as one-dimensional (1D), two-dimensional (2D), or multiple-dimensional (nD). There are two types of data structure: homogeneous and heterogeneous. The homogeneous data structure allows to store the identical type of data. The heterogeneous data structure that allows to store the elements are often various types also. The most common data structure in R are vector, factor, matrix, array, list and dataframe as shown in Figure 7.1.

Vector is the basic data structure in R. It is a one-dimensional and homogeneous data structures. There are six types of atomic vectors such as integer, character, logical, double or raw. It is a collection of elements, which is most commonly of mode character, inter, logical, or numeric [1, 2].

Factor is a data object, which is used to categorize the data and store it as a level. It can store both integers and strings. It has two attributes, such as

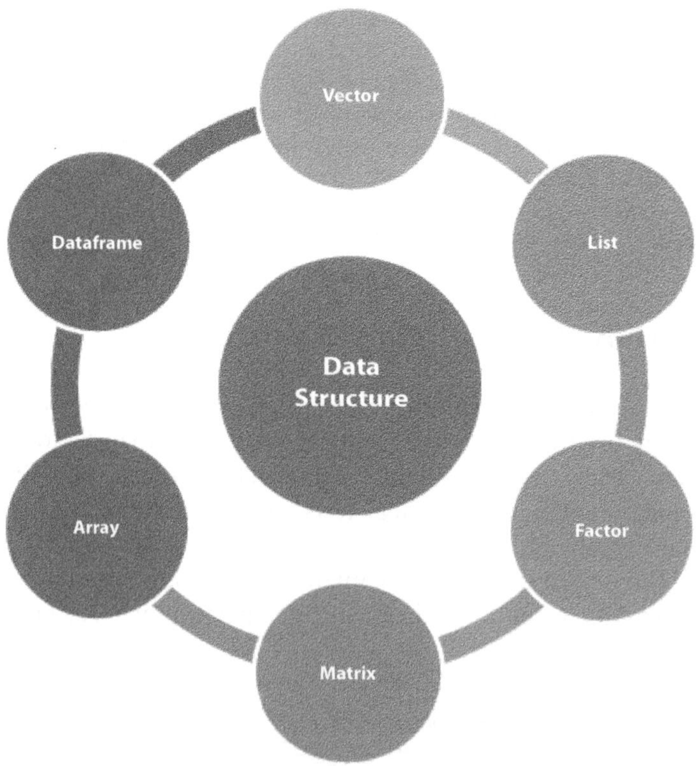

Figure 7.1 Data structure in R.

Table 7.1 Classified view of data structures in R.

Data types	**Same data type**	**Multiple data type**
Number of dimensions		
One	Vector	List
One (Categorical data)	Factor	
Two	Matrix	Data Frame
Many	Array	

class and level, where class has a value of factor, and level is a set of allowed values (refer to Figure 7.1).

Matrix is a two-dimensional and homogeneous data structures. All the values in a matrix have a same data type. It is a rectangular arrangement of rows and columns.

Array is a three-dimensional or more to store the data. It is a homogeneous data structure. It is a collection of a similar data types with continues memory allocation.

List is the collection of data structure. It is a heterogeneous data structure. It is very similar to vectors except they can store data of different types of mixture of data types. It is a special type of vector in which each element can be a different data type. It is a much more complicated structure.

Data frame is a two-dimensional and heterogeneous data structures. It is used to store the data object in tabular format in rows and columns.

These data structures are further classified into the following way on the basis of on the types of data and number of dimensions as shown in Table 7.1.

Data structures are classified based on the types of data that they can hold like homogeneous and heterogeneous. Now let us discuss all the data structures in detail with its characteristics and examples.

7.2 Homogeneous Data Structures

The data structures, which can hold the similar type of data, can be referred as homogeneous data structures.

7.2.1 Vector

Vector is a basic data structure in R. The vector may contain single element or multiple elements. The single element vector with six different types of atomic vectors, such as integer, double, character, logical, complex, and raw are as below:

```
# Integer type of atomic vector
print(25L)
[1] 25

# Double type of atomic vector
print(83.6)
[1] 83.6

# Character type of atomic vector
print("R-Programming")
[1] "R-Programming"

# Logical type of atomic vector
print(FALSE)
[1] FALSE

# Complex type of atomic vector
print(5+2i)
[1] 5+2i

# Raw type of atomic vector
print(charToRaw("Test"))
[1] 54 65 73 74
```

- **Using Colon (:) Operator**

The following examples will create vectors using colon operator as follows:

```
# Create a series from 51 to 60
vec <- 51:60
print(vec)
[1] 51 52 53 54 55 56 57 58 59 60

# Create a series from 5.5 to 9.5
vec <- 5.5:9.5
print(vec)
[1] 5.5 6.5 7.5 8.5 9.5
```

- **Using Sequence (seq) Operator**

The following examples will create vectors using sequence operator as follows:

```
# Create a vector from 1 to 10 incremented
  by 2
print(seq(1, 10, by=2))
[1] 1 3 5 7 9

# Create a vector from 1 to 50 incremented
  by 5
print(seq(1, 50, by=5))
[1]  1  6 11 16 21 26 31 36 41 46

# Create a vector from 5 to 6 incremented
  by 0.2
print(seq(5,6, by=0.2))
[1] 5.0 5.2 5.4 5.6 5.8 6.0
```

- **Using c() Function**

The vector can be created using ***c()*** function for more than one element in a single vector. It combines the different elements into a vector. The following code will create a simple vector named as ***color*** with Red, Green, Blue, Pink and Yellow as an element.

```
# Create a vector
color <- c("Red", "Green", "Blue", "Pink",
  "Yellow")
print(color)
[1] "Red"       "Green"    "Blue"      "Pink"
  "Yellow"
```

The ***class()*** function is used to find the class of elements of vector. The following code will display the class of vector ***color***.

```
# Class of a vector
print(class(color))
[1] "character"
```

The non-character values in a vector are converted into character type as follows.

```
# Numeric value is converted into characters
char <- c("Color", "Purple", 10)
print(char)
[1] "Color"  "Purple" "10"
```

- **Accessing Vector Elements**

The elements of vector can be access using index. The [ ] bracket is used for indexing. The index value is start from 1. The below code will display the third, seventh and ninth elements of a vector ***month***.

```
# Using position
mon <- c("JAN", "FEB", "MAR", "APR", "MAY",
  "JUN", "JUL", "AUG", "SEP", "OCT", "NOV",
  "DEC")
res <- mon[c(3,7,9)]
print(res)
[1] "MAR" "JUL" "SEP"
```

The vector elements can be access using logical indexing also. The below code will display the first, fourth and sixth elements of a vector month.

```
# Using logical index
mon <- c("JAN", "FEB", "MAR", "APR", "MAY",
  "JUN")
res <- mon[c(FALSE,TRUE,FALSE,TRUE,FALSE,
  TRUE)]
print(res)
[1] "FEB" "APR" "JUN"
```

The vector elements can be access using negative indexing also. The negative index value is skipped. The below code will skip third and sixth elements of a vector month.

```
# Using negative index
mon <- c("JAN", "FEB", "MAR", "APR", "MAY",
  "JUN")
res <- mon[c(-3,-6)]
print(res)
[1] "JAN" "FEB" "APR" "MAY"
```

The vector elements can be access using 0/1 indexing also. The below code will display first and fourth elements of a vector month.

```
# Using 0 and 1 index
mon <- c("JAN", "FEB", "MAR", "APR", "MAY",
  "JUN")
res <- mon[c(1,0,0,0,4,0)]
print(res)
[1] "JAN" "APR"
```

• **Nesting of Vectors**
The multiple vectors can be combined together to create a vector is called nesting of vectors. We can combine two or more vectors to create a new vector or we can use a vector with other values to create a vector.

```
# Creating a vector from two vectors
vec1 <- c(21,22,23)
vec2 <- c(24,25,26)
vec3 <- c(vec1,vec2)
print(vec3)
[1] 21 22 23 24 25 26

# Adding more values in a vector
vec4 <- c(vec3,27,28,29)
print(vec4)
[1] 21 22 23 24 25 26 27 28 29

# Creating a vector from three vectors
vec5 <- c(vec3,vec2,vec1)
print(vect5)
[1] 21 22 23 24 25 26 24 25 26 21 22 23
```

• **Vector Arithmetic**
The various arithmetic operations can be performed on two or more same length of vectors. The operation can be addition, subtraction, multiplication or division as follows:

```
# Create vectors
vec1 <- c(8,5,7,8,9,2,3,5,1)
vec2 <- c(5,7,3,6,8,2,4,6,0)
```

```
# Vector addition
add = vec1+vec2
print(add)
[1] 13 12 10 14 17  4  7 11  1

# Vector subtraction
sub = vec1-vec2
print(sub)
[1]  3 -2  4  2  1  0 -1 -1  1

# Vector multiplication
mul = vec1*vec2
print(mul)
[1] 40 35 21 48 72  4 12 30  0

# Vector division
div = vec1/vec2
print(div)
[1] 1.6000000 0.7142857 2.3333333 1.3333333
  1.1250000
[6] 1.0000000 0.7500000 0.8333333       Inf
```

- **Vector Element Recycling**

The various operations can be performed on vectors of different length also. The elements of a shorter vectors are recycled to complete the operations as follows:

```
# Create vector
vec1 <- c(6,3,7,5,9,1,6,5,2)
vec2 <- c(4,7,2)     # here v2 becomes
  c(4,7,2,4,7,2,4,7,2)
print(vec1+vec2)
[1] 10 10  9  9 16  3 10 12  4
```

- **Sorting of Vector**

The elements of a vector can be sorting (ascending / descending) using ***sort()*** function.

The below code will display elements of a vector in ascending order as follows:

```
# Sorting vector
vec1 <- c(45,12,8,56,-23,71)
```

```
res <- sort(vec1)
print(res)
[1] -23   8  12  45  56  71

# Sorting character vector
fruit <- c("Banana", "Apple", "Mango",
  "Orange", "Grapes", "Kiwi")
res <- sort(fruit)
print(res)
[1] "Apple"   "Banana"  "Grapes"  "Kiwi"
  "Mango"  "Orange"
```

The below code will display elements of a vector in descending order as follows:

```
# Sorting vector in descending order
vec1 <- c(45,12,8,56,-23,71)
res <- sort(vec1, decreasing = TRUE)
print(res)
[1]  71  56  45  12   8 -23

# Sorting character vector in descending
  order
fruit <- c("Banana", "Apple", "Mango",
  "Orange", "Grapes", "Kiwi")
res <- sort(fruit, decr=TRUE)
print(res)
[1] "Orange" "Mango"  "Kiwi"   "Grapes"
  "Banana" "Apple"
```

7.2.2 Factor

The factor is used to categorized the data and store it as levels. It has a limited number of unique values. It is useful in data analysis for statistical modelling. The ***factor()*** function is used to create factors.

The following example will create a vector ***bg*** and apply factor function to convert the vector into a factor. It will display as follows:

```
# Create a vector
bg <- c("A","A","O","O","AB","A","A","B")
```

```
# Apply factor function to a vector and
  print it
factor_bg <- factor(bg)
print(factor_bg)
[1] A  A  O  O  AB A  A  B
Levels: A AB B O
```

The above code creates into four levels.
The structure of factor is display using ***str()*** function as follows

```
# Structure of a factor function
str(factor_bg)
Factor w/ 4 levels "A","AB","B","O": 1 1 4
  4 2 1 1 3
```

It is a level of factor, which is an alphabetical order and it can observe that for each level of an integer is assigned into the factor, which can save the memory space.

7.2.3 Matrix

Matrix is a data structure in which the elements are arranged in a two-dimensional format. All the elements in a metrices of the same atomic types. The numeric elements of matrices are to be used for mathematical calculation. The matrix can be created using ***matrix()*** function as follows

```
matrix(data, nrow, ncol, byrow, dimnames)
```

Here,
data – An input vector
nrow – No. of rows
ncol – No. of columns
byrow – TRUE or FALSE
dimname – Name of rows and columns

• **Create Matrix**
The following example will create a numeric matrix.

```
# Create a row wise matrix
MAT1 <- matrix(c(21:29), nrow = 3, byrow =
  TRUE)
print(MAT1)
```

```
      [,1] [,2] [,3]
[1,]   21   22   23
[2,]   24   25   26
[3,]   27   28   29
```

In above example, it is set to create three rows and display the matrix row wise.

The following example will create a numeric matrix.

```
# Create a column wise matrix
MAT2 <- matrix(c(31:39), nrow = 3, byrow =
  FALSE)
print(MAT2)
      [,1] [,2] [,3]
[1,]   31   34   37
[2,]   32   35   38
[3,]   33   36   39
```

In above example, it is set to create three rows and display the matrix column wise.

- **Assigning Rows and Columns Names**

The following example will assign the names of rows and columns and creates a numeric matrix.

```
# Assigning the name of rows and columns
rname = c("Row1","Row2","Row3")
cname = c("Col1","Col2","Col3")

# Create and print a matrix with its rows
  and column names
MAT <- matrix(c(41:49), nrow=3, byrow=TRUE,
  dimnames = list(rname,cname))
print(MAT)
     Col1 Col2 Col3
Row1   41   42   43
Row2   44   45   46
Row3   47   48   49
```

In above example, it is assigned row names such as Row1, Row2, and Row3 and columns names such as Col1, Col2, and Col3. It is also set to create three rows and display all the elements in a row wise in a matrix.

- **Assessing Matrix Elements**

The matrix elements can be accessed by combination of row and column index. The following example will access the matrix elements as follows:

```
# Accessing the element at 1st row and 3rd
  column
print(MAT[1,3])
[1] 43

# Accessing the element at 2nd row and 2nd
  column
print(MAT[2,2])
[1] 45

# Accessing all the elements of 3rd row
print(MAT[3,])
Col1 Col2 Col3
  47   48   49
# Accessing all the elements of 1st Column
print(MAT[,1])
Row1 Row2 Row3
  41   44   47
```

- **Updating Matrix Elements**

We can assign a new value to the element of a matrix using its location of the elements. The following example will update the value of matrix element as follows:

```
# Create and print matrix
MAT <- matrix(c(21:29), nrow = 3, byrow =
  TRUE)
print(MAT)
     [,1] [,2] [,3]
[1,]   21   22   23
[2,]   24   25   26
[3,]   27   28   29

# Accessing the 2nd row and 2nd column
  element
MAT[2,2]
[1] 25
```

```
# Update the M[2,2] value with 99
MAT[2,2]<-99
print(MAT)
     [,1] [,2] [,3]
[1,]   21   22   23
[2,]   24   99   26
[3,]   27   28   29
```

• **Matrix Computation**

The various arithmetic operation can be performed on a matrix. The result of the operations is also stored in a matrix. The following examples will perform the various operation such as matrix addition, subtraction, multiplication and division.

```
# Matrix Addition
mat_add <- MAT1 + MAT2
print(mat_add)
     [,1] [,2] [,3]
[1,]   52   56   60
[2,]   56   60   64
[3,]   60   64   68

# Matrix Subtraction
mat_sub <- MAT2 - MAT1
print(mat_sub)
     [,1] [,2] [,3]
[1,]   10   12   14
[2,]    8   10   12
[3,]    6    8   10

# Matrix Multiplication
mat_mul <- MAT1 * MAT2
print(mat_mul)
     [,1] [,2] [,3]
[1,]  651  748  851
[2,]  768  875  988
[3,]  891 1008 1131

# Matrix Division
mat_div <- MAT2 / MAT1
print(mat_div)
```

```
            [,1]     [,2]     [,3]
[1,] 1.476190 1.545455 1.608696
[2,] 1.333333 1.400000 1.461538
[3,] 1.222222 1.285714 1.344828
```

- **Transpose of Matrix**

Transposition is a process to swapped the rows and columns with each other's in a matrix. The ***t()*** function is used to find the transpose of a given matrix. The following example will find the transpose matrix of an input matrix as follows:

```
# Create matrix
MAT <- matrix(c(21:29), nrow = 3, byrow =
  TRUE)

# Print matrix
print(MAT)
     [,1] [,2] [,3]
[1,]   21   22   23
[2,]   24   25   26
[3,]   27   28   29

# Print transpose of a matrix
print(t(MAT))
     [,1] [,2] [,3]
[1,]   21   24   27
[2,]   22   25   28
[3,]   23   26   29
```

7.2.4 Array

Array can be store the data in two or more dimensions also. The array can be created using ***array()*** function. The vector is used as an input and dim parameter is used to create an array.

The following example will create an array of two 3X3 matrices with three rows and three columns as follows:

```
# Create vectors
x1 <- c(11,12,13)
x2 <- c(14,15,16,17,18,19)
```

```
# Create array using vectors
x <- array(c(x1,x2),c(3,3,2))
print(x)
, , 1

     [,1] [,2] [,3]
[1,]   11   14   17
[2,]   12   15   18
[3,]   13   16   19

, , 2

     [,1] [,2] [,3]
[1,]   11   14   17
[2,]   12   15   18
[3,]   13   16   19
```

The following example will create an array of four 2 × 2 matrices with two rows and two columns as follows:

```
# Create array
x <- array(c(1:16),c(2,2,4))
print(x)
, , 1

     [,1] [,2]
[1,]    1    3
[2,]    2    4

, , 2

     [,1] [,2]
[1,]    5    7
[2,]    6    8

, , 3

     [,1] [,2]
[1,]    9   11
[2,]   10   12
```

```
, , 4

     [,1] [,2]
[1,]   13   15
[2,]   14   16
```

The name of rows, columns, and matrix is also to be assigned as follows:

```
# Assigning the name of rows, columns and
  matrix
rname <- c("ROW1","ROW2","ROW3")
cname <- c("COL1","COL2","COL3")
mname <- c("Matrix-1","Matrix-2")

# Create and print a matrix with its names
x <- array(c(21:38), c(3,3,2), dimnames =
  list(cname,rname,mname))
print(x)
, , Matrix-1

     ROW1 ROW2 ROW3
COL1   21   24   27
COL2   22   25   28
COL3   23   26   29

, , Matrix-2

     ROW1 ROW2 ROW3
COL1   30   33   36
COL2   31   34   37
COL3   32   35   38
```

7.3 Heterogeneous Data Structures

The data structure, which is capable of storing different types of data, is referred as heterogeneous data structures. As mentioned in Table 7.1, R is supporting list and data frame for holding different types of data in one dimensional or multidimensional format.

7.3.1 List

It is a data structure that consists various types of elements in a list, such as numeric, string, vector, list, etc.

• **Create List**
The list can be created using ***list()*** function. The following example will create a list ***lst*** using various types of elements inside it.

```
# Create and print a list
lst <- list("Banana", c(50,78,92), TRUE,
  83.68)
print(lst)

[[1]]
[1] "Banana"

[[2]]
[1] 50 78 92

[[3]]
[1] TRUE

[[4]]
[1] 83.68
```

The above list contains the four different types of elements such as character, vector, logical and numeric.

• **Naming List Elements**
We can assign a name of each elements in a list. The name will be used to access each elements of a list separately. The following example will create a list ***lst*** using matrix, vector, and list inside it.

```
# Create a list
lst <- list(matrix(c(11,12,13,14,15,16,17,
  18,19), nrow=3), c("Saturday", "Sunday"),
  list("Banana",83.68))

# Naming of elements in a list
names(lst) <- c("Matrix", "Weekend", "List")
print(lst)
```

```
$Matrix
     [,1] [,2] [,3]
[1,]   11   14   17
[2,]   12   15   18
[3,]   13   16   19

$Weekend
[1] "Saturday" "Sunday"

$List
$List[[1]]
[1] "Banana"

$List[[2]]
[1] 83.68
```

The above example assigns a name Matrix, Weekend, and List to the elements of list.

- **Accessing List Elements**

The following examples will be accessing the elements of list using indexing.

```
# Accessing 1st element of a list
print(lst[1])
$Matrix
     [,1] [,2] [,3]
[1,]   11   14   17
[2,]   12   15   18
[3,]   13   16   19

# Accessing 2nd element of a list
print(lst[2])
$Weekend
[1] "Saturday" "Sunday"

# Accessing 3rd element of a list
print(lst[3])
$List
$List[[1]]
[1] "Banana"
```

```
$List[[2]]
[1] 83.68
```

The following examples will be accessing the elements of list using its names.

```
# Accessing list element using its name
print(lst$Matrix)
     [,1] [,2] [,3]
[1,]   11   14   17
[2,]   12   15   18
[3,]   13   16   19

# Accessing list element using its name
print(lst$Weekend)
[1] "Saturday" "Sunday"

# Accessing list element using its name
print(lst$List)
[[1]]
[1] "Banana"

[[2]]
[1] 83.68
```

The ***length()*** function is used to find the length of a list, the ***str()*** function is used to display the structure of a list and the ***summary()*** function is used to display the summary of a list.

The following examples will find the length of a list, display the structure and summary of a list.

```
# Find the length of a list
length(lst)
[1] 3

# Display the structure of a list
str(lst)
List of 3
 $ : num [1:3, 1:3] 11 12 13 14 15 16 17
  18 19
 $ : chr [1:2] "Saturday" "Sunday"
```

```
 $ :List of 2
  ..$ : chr "Banana"
  ..$ : num 83.7

# Display the summary of a list
summary(lst)
     Length Class  Mode
[1,] 9      -none- numeric
[2,] 2      -none- character
[3,] 2      -none- list
```

- **Manipulating Elements of List**

The elements in a list will be manipulated using addition of new elements in a list, deleting elements from the list and update the elements in a list. The following example will show the add, delete, and update operation in a list.

```
# Add a new element in a list
lst[4]<- "Orange"
print(lst[4])
[[1]]
[1] "Orange"

# Update the fourth element of the list
lst[4]<- "Red"
print(lst[4])
[[1]]
[1] "Red"

# Delete the element in a list
lst[4] <- NULL
print(lst[4])
$<NA>
NULL
```

- **Merging List Elements**

The two or more list can be merge into a single list with its all elements. The following example will create two lists, such as ***lst1*** and ***lst2***. The both lists will merge into a single list as follows:

```
# Create list1
lst1 <- list(1,2,3,4,5)
```

```
# Create list2
lst2 <- list("Six", "Seven", "Eight",
  "Nine", "Ten")

# Merging list1 and list2
lst <- c(lst1,lst2)

# Display the final merge list
print(lst)

[[1]]
[1] 1

[[2]]
[1] 2

[[3]]
[1] 3

[[4]]
[1] 4

[[5]]
[1] 5

[[6]]
[1] "Six"

[[7]]
[1] "Seven"

[[8]]
[1] "Eight"

[[9]]
[1] "Nine"

[[10]]
[1] "Ten"
```

7.3.2 Dataframe

The dataframe is a table-like structure. It is a fundamental data structure to store these types of dataset in which data is organized in number of observations and number of variables. In data frame multiple types of data is stored in multiple labeled columns and it is a prime difference between matrix and data frame. Elements of same column should of same type is an observable restriction in data frame.

The dataframe can be imported from the various sources, like CSV file, excel file, SPSS, relational database etc. The dataframe can be created manually also.

• Create Dataframe

The ***data.frame()*** function is used to create a dataframe manually.

The following example will create a ***stud*** dataframe with column names Rno, Name and City.

```
# Create vectors
Rno = c(101,102,103,104,105)
Name = c("Rajan", "Vraj", "Manshi", "Jay",
  "Tulsi")
City = c("Rajkot","Baroda","Surat","Ahmed-
  abad","Valsad")

# Create data frames
stud = data.frame(Rno, Name, City)
print(stud)
  Rno   Name      City
1 101  Rajan    Rajkot
2 102   Vraj    Baroda
3 103 Manshi     Surat
4 104    Jay Ahmedabad
5 105  Tulsi    Valsad
```

• Addition of Column

We can add a new column in the existing data frame. The following example will add a new column ***Age*** in the stud data frame as follows:

```
# Create vector
Age = c(23,26,24,25,24)
```

```
# Add new column into a data frame
stud = data.frame(Rno, Name, City, Age)
print(stud)
  Rno   Name      City Age
1 101  Rajan    Rajkot  23
2 102   Vraj    Baroda  26
3 103 Manshi     Surat  24
4 104    Jay Ahmedabad  25
5 105  Tulsi    Valsad  24
```

• **Accessing Dataframe**

The dataframe can be access as follows:

```
# Display 1st row
stud[1,]
  Rno  Name   City Age
1 101 Rajan Rajkot  23

# Display 2nd Column
stud[2]
    Name
1  Rajan
2   Vraj
3 Manshi
4    Jay
5  Tulsi

# Display 2nd and 3rd row with only selected
  column
stud[c(2,3),c("Name","City")]
    Name   City
2   Vraj Baroda
3   Manshi  Surat
```

R provides an interactive environment for data analysis and statistical computing. It supports several basic various data types that are frequently used in different calculation and analysis-related work. It supports six basic data types, such as numeric (real or decimal), integer, character, logical, complex, and raw.

References

1. Bercea, I.O. Even, G., An extendable data structure for incremental stable perfect hashing, in: *STOC 2022 - Proceedings of the 54th Annual ACM SIGACT Symposium on Theory of Computing. (Proceedings of the Annual ACM Symposium on Theory of Computing)*. S. Leonardi, & A. Gupta (Eds.), pp. 1298–1310, Association for Computing Machinery, 2022. https://doi.org/10.1145/3519935.3520070.
2. Ozturk, Z., Topcuoglu, H. R., Kandemir, M.T., Studying error propagation on application data structure and hardware. *Journal of Supercomput.*, 78, 17, 18691–18724, 2022. https://doi.org/10.1007/s11227-022-04625-x
3. Wickham, H. and Grolemund, G., R for data science: Import, tidy, transform, visualize, and model data, Paperback – 4 February 2017.
4. Prakash, P.K.S., Krishna Rao, A.S., *R data structures and algorithms.* Packt Publishing; 1st edition, 21 November 2016.

8

Dimension Reduction Techniques in Distributional Semantics: An Application Specific Review

Pooja Kherwa[1*], Jyoti Khurana[2], Rahul Budhraj[1], Sakshi Gill[1], Shreyansh Sharma[1] and Sonia Rathee[1]

[1]Department of Computer Science, Maharaja Surajmal Institute of Technology, New Delhi, India
[2]Department of Information Technology, Maharaja Surajmal Institute of Technology, New Delhi, India

Abstract
In recent years, the data tends to be very large and complex and it becomes very difficult and tedious to work with large datasets containing huge number of features. That's where Dimensionality Reduction comes into play. Dimensionality Reduction is a pre-processing step in various fields such as machine learning, data mining, statistics etc. and is effective in removing irrelevant and highly redundant data. In this paper, the author's performed a vast literature survey and aims to provide an adequate application based understanding of various dimensionality reduction techniques and to work as a guide to choose right approach of Dimensionality Reduction for better performance in different applications. Here, the authors have also performed detailed experiments on two different datasets for comparative analysis between various linear and non-linear dimensionality reduction techniques to figure out the effectiveness of the techniques used. PCA, a linear dimensionality reduction technique, outperformed all other techniques used in the experiments. In fact, almost all the linear dimensionality reduction techniques outperformed the non-linear techniques on both datasets by a huge error percentage margin.

***Keywords*:** Dimension reduction, principal component analysis, single value decomposition, auto encoders, factor analysis

**Corresponding author*: poojakherwa@gmail.com

M. Niranjanamurthy, Kavita Sheoran, Geetika Dhand, and Prabhjot Kaur (eds.) Data Wrangling: Concepts, Applications and Tools, (147–186) © 2023 Scrivener Publishing LLC

8.1 Introduction

Dimensionality Reduction is a pre-processing step, which aims at reducing the original high dimensionality of a dataset, to its intrinsic dimensionality. Intrinsic Dimensionality of a dataset is the minimum number of dimensions, or variables, in which the data can be represented without suffering any loss. So far in this field, achieving intrinsic dimensionality is a near ideal situation. With years and years of handwork, the brightest mind in this field, have achieved up to 97% of their goal, but it could not be 100%. So, it won't be wrong of us to say that we are still in development phase of this field.

Domains such as Machine Learning, Data Mining, Numerical Analysis, Sampling, Combinatorics, Databases, etc., suffer from a very popular phenomenon, called "The Curse of Dimensionality". It refers to the issues that occur while analysing and organising data in high dimensional spaces. The only way to deal with it is Dimensionality reduction. Not only this, it helps to avoid Over fitting, which occurs when noise is captured by a model, or an algorithm. Dimensionality Reduction removes redundant information and leads to an improved classifier accuracy.

The transition of dataset representation from a high-dimensional space to a low-dimensional one can be done by using two different approaches, i.e., Feature Selection methods and Feature extraction methods. While the former approach basically selects the more suitable features/parameters/variables, for the low dimensional subspace, from the original set of parameters, the latter assists the mapping from high dimensional input space to the low dimensional target space by extracting a new set of parameters, from the existing set [18]. Mohini D. Patil & Shirish S. Sane [4], have presented a brief review on both the approaches in their paper. Another division of techniques can be done on the basis of nature of datasets, namely, Linear Dimension Reduction techniques and Non-Linear Dimension Reduction techniques. As the names suggest, linear techniques are applied on linear datasets, whereas Non-linear techniques work for Non-Linear datasets. Principal Component Analysis (PCA) is a traditional technique, which has achieved peaks of success over the past few decades. But being a Linear Dimension Reduction technique, it is an incompetent algorithm for complex and non-linear datasets. The recent invasion in the technological field over the past few years, has led to generation of more complex data, with a nature of non-linearity. Hence, the focus has now shifted to Non-Linear Dimension Reduction algorithms. In [24], L.J.P. van der Maaten *et al.* put forth a detailed comparative review of 12 non-linear techniques, which included performing experiments on natural, as well as artificial datasets. Joshua B. Tenenbaum *et al.* [20] have described a non-linear approach that combines the major algorithmic features of PCA and MDS. There exists one more way for the classification of Dimension Reduction approaches, Supervised & Unsupervised approaches. Supervised techniques make use of class information, for example: LDA, Neural Networks, etc. Whereas unsupervised techniques don't use any label information. Clustering is an example of unsupervised approach.

Figure 8.1 depicts a block diagram of the process of Dimensionality Reduction.

Computer science is a very vast domain, and the data generated in it is incomparable. Dimensionality Reduction has played a crucial role for data compression in this domain for decades now. From statistics to machine learning, its applications have been increasing with a tremendous rate. Facial recognition, MRI scans, image processing, Neuroscience, Agriculture, Security applications, E-commerce, Research work, Social Media, etc. are just a few examples of its application areas. Such development, which we are witnessing right now, owes a great part of their success to this phenomenon. Different approaches are applied for different applications, based on the advantages and drawbacks of the approach and the demands of the datasets. Expecting one technique to satisfy the needs of all the datasets is not justified.

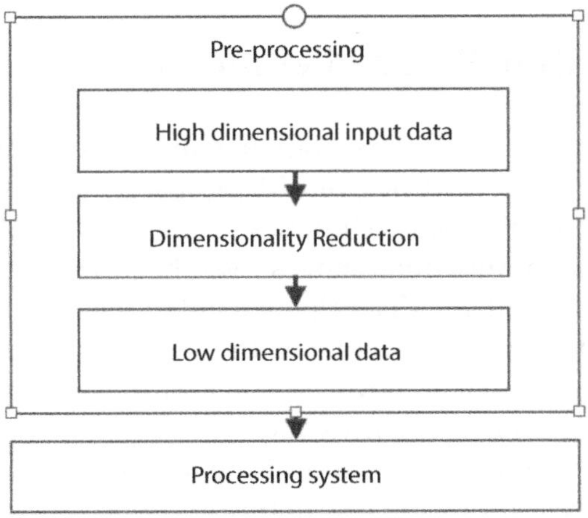

Figure 8.1 Overview of procedure of dimensionality reduction.

The studies we have surveyed so far focus on either providing a generalised review of various techniques, such as, Alireza Sarveniaza [2] provided a review of various linear and non-linear dimension reduction methods, or a comparative review of techniques based on a few datasets like, Christoph Bartenhagen *et al.* [3] did a study which compared various unsupervised techniques, on the basis of their performance on micro-array data. But our study provides a detailed comparative review of techniques based on application areas, which would prove to be helpful for deciding the suitable techniques for datasets based on their nature. This paper aims at serving as a guide for providing apt suggestions to researchers and computer science enthusiasts, when struggling to choose between various Dimensionality Reduction techniques, so as to yield a better result. The flow of the paper is as follows: (Section 8.1 provides Introduction to Dimensionality Reduction,) Section 8.2 classifies Dimension Reduction Techniques on the basis of applications. Section 8.3 reviews 10 techniques namely, Principal Component Analysis (PCA), Linear Discriminant Analysis (LDA), Kernel Principal Component Analysis (KPCA), Locally Linear Embedding (LLE), Independent Component Analysis (ICA), Isomap (IM), Self-Organising Map (SOM), Singular Value Decomposition (SVD), Factor Analysis (FA) and Auto-Encoders. Section 8.4 provides a detailed summary of the observations and the factors affecting the performance of the Dimensionality Reduction techniques on two natural datasets. Section 8.5 lists out the results of the experiments. It also represents the basic analysis of the experimental survey. Section 8.6 concludes the paper and section 8.7 lists out the references used for this survey.

8.2 Application Based Literature Review

Figure 8.2 is a summed-up representation of the usage of dimension reduction in the three fields, Statistics, Bio-Medical and Data Mining, and a list of the most commonly used techniques in these fields. The diagram is a result of the research work done and it serves the primary goal of the paper, i.e., it provides the reader with a helping hand to select suitable techniques for performing dimension reduction on the datasets, based on their nature. The techniques mentioned above are some of the bests performing and most used techniques in these fields. It can be clearly seen that Bio-medical is the most explored field. More work is being done in the Statistics field. For a detailed overview of the research work done for this paper, and in order to gain more perspective regarding the usage of various tools for different applications, Table 8.1 has been formed. Most of the papers referred

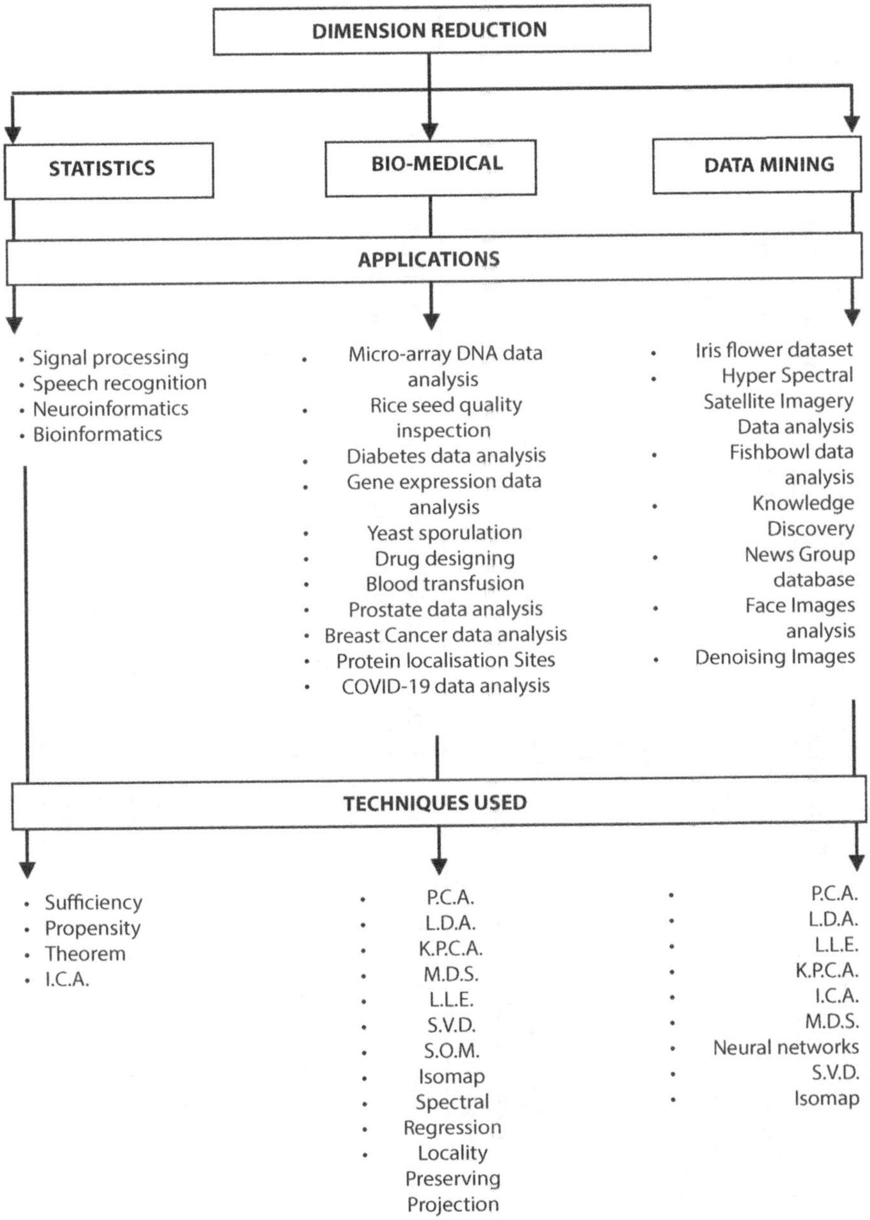

Figure 8.2 Dimension reduction techniques and their application areas.

Table 8.1 Research papers and the tools and application areas covered by them.

S. no.	Paper name	Tools/techniques	Application areas
1	Experimental Survey of Various DR Techniques [1]	Standard Deviation, Variance, PCA, LDA, Factor Analysis	Data Mining, Iris-Flower Dataset
2	An Actual Survey of DR [2]	PCA, KPCA, LDA, CCA, OPCA, NN, MDS, LLE, IM, EM, Principal Curves, Nystroem, Graph-based and new methods	3
3	Comparative Study of Unsupervised DR for Visualization of Microarray Gene Expression [3]	PCA, KPCA, IM, MVU, DM, LLE, LEM	Microarray DNA Data
4	Dimension Reduction: A Review [4]	Feature Selection algos, Feature Extraction algos: LDA, PCA [combined algos proposed]	Data Mining, Knowledge Discovery
5	Most Informative Dimension Reduction [5]	Iterative Projection algo	Statistics, Document Categorization, Bio-Informatics, Neural Code Analyser
6	Sufficient DR Summaries [6]	Sufficiency, Propensity Theorem	Statistics
7	A Review on Dimension Reduction [7]	Inverse Regression based methods, Non-parametric and semi parametric methods, inference	Statistics

(Continued)

Table 8.1 Research papers and the tools and application areas covered by them. (*Continued*)

S. no.	Paper name	Tools/techniques	Application areas
8	Global versus Local Methods in Non-Linear DR [8]	MDS, LLE,(Conformal IM, Landmark IM) Extensions of IM	Fishbowl Dataset, Face Images Dataset
9	A Review on DR in Data Mining [9]	ICA, KPCA, LDA, NN, PCA, SVD	Data Mining
10	Comparative Study of PCA & LDA for Rice Seeds Quality Inspection [10]	PCA, LDA, Random Forest Classifier, Hyper-spectral Imaging	Rice Seed Quality inspection
11	Sparse KPCA [11]	KPCA, Max. Likelihood Approach	Diabetes Dataset, 7-D Prima Indians, Non-Linear Problems
12	Face Recognition using KPCA [12]	KPCA	Face Recognition, Face Processing
13	Sparse KPCA for Feature Extraction in Speech Recognition [13]	KPCA, PCA, Maximum Likelihood	Speech Recognition
14	PCA for Clustering Gene Expression Data [14]	Clustering algos (CAST, K-Means, Average Link), PCA	Gene expression Data, Bio-Informatics
15	PCA to Summarize Microarray Expressions [15]	PCA	DNA Microarray data, Bio-Informatics, Yeast Sporulation
16	Reducing Dimension of Data with Neural Network [16]	Deep Neural Networks, PCA, RBM	Handwritten Digits Datasets

(*Continued*)

Table 8.1 Research papers and the tools and application areas covered by them. (*Continued*)

S. no.	Paper name	Tools/techniques	Application areas
17	Robust KPCA [17]	KPCA, Novel Cost Function	Denoising Images, Intra-Sample Outliers, Find missing data, Visual data
18	Dimensionality Reduction using Genetic Algos [18]	GA Feature Extractor, KNN, Sequence Floating Forward Sel.	Biochemistry, Drug Design, Pattern Recognition
19	Non Linear Dimensionality Reduction [19]	Auto-Association Technique, Greedy Algo, Encoder, Decoder	Time Series, Face Images, Circle & Helix problem
20	A Global Geometric Framework for NLDR [20]	Isomap, (PCA + MDS)	Vision, Speech, Motor Control, Physical & Biological Sciences
21	Semi-Supervised Dimension Reduction [21]	KNN Classifier, PCA, cFLD, SSDR-M, SSDR-CM, SSDR-CMU	Data Mining, UCI Dataset, Face Images, News Group Database
22	Application of DR in Recommender System: A Case Study [22]	Collaborative Filtering, SVD, KDD, LSI Technique	E-Commerce, Knowledge Discovery Database
23	Classification Constrained Dimension Reduction [23]	CCDR Algo, KNN< PCA, MDS, IM, Fischer Analysis	Label Info, Data Mining, Hyper Spectral Satellite Imagery Data

(*Continued*)

Table 8.1 Research papers and the tools and application areas covered by them. (*Continued*)

S. no.	Paper name	Tools/techniques	Application areas
24	Dimensionality Reduction: A Comparative Review [24]	PCA, MDS, IM, MVU, KPCA, Multilayer Auto Encoders, DM, LLE, LEM, Hessian LLE, Local Tangent Space Analysis, Manifold Charting, Locally Linear Coordination	DR, Feature Extraction, Manifold Learning, Handwritten Digits, Pedestrian Detection, Face Recognition, Drug Discovery, Artificial Datasets
25	Sufficient DR & Prediction in Regression [25]	SDR, Regression, PCs, New Method designed for Prediction, Inverse Regression Models	Sufficient Dimension Reduction
26	Hyperparameter Selection in KPCA [26]	KPCA	
27	KPCA and its Applications in Face Recognition and Active Shape Models [27]	KPCA	Pattern Classification, Face Recognition
28	Validation Study of DR Impact on Breast Cancer Classification [28]	LLE, IM, Locality Preserving Projection (LPP), Spectral Regression (SR)	Breast Cancer Data
29	Dimensionality Reduction[29]	PCA, IM, LLE	Time series data analysis
30	Dimension Reduction of Health Data Clustering [30]	SVD, PCA, SOM, ICA	Acute Implant, Blood Transfusion, Prostate Cancer

(*Continued*)

Table 8.1 Research papers and the tools and application areas covered by them. (*Continued*)

S. no.	Paper name	Tools/techniques	Application areas
31	The Role of DR in Classification [31]	RBF Mapping with a Linear SVM	MNIST-10 Classes, K-Spiral Dataset
32	Dimension reduction [32]	PCA, LDA, LSA, Feature Selection Techniques: Filter, Wrapper, Embedded approach	Importance of DR
33	Fast DR and Simple PCA [33]	PCA	Handwritten Digits in English & Japanese Kanji
34	Comparative Analysis of DR in ML [34]	LDA, PCA, KPCA	Iris Dataset (Plants), Wine
35	A Survey of DR and Classification methods [35]	SVD, PCA, ICA, CCA, LLE, LDA, PLS Regression	General Importance of DR in Data Processing
36	A Survey of DR Techniques [36]	PCA, SVD, Non-Linear PCA, Self-Organising Maps, KPCA, GTM, Factor Analysis	General Importance of these techniques
37	Non-Linear DR by LLE [37]	LLE	Face Images, Vectors of Word Document
38	Survey on Feature Selection & DR Techniques [38]	SVD, PLSR, LLE, PCA, ICA, CCA	Data Mining

(*Continued*)

Table 8.1 Research papers and the tools and application areas covered by them. (*Continued*)

S. no.	Paper name	Tools/techniques	Application areas
39	Alternative Model for Extracting Multi-dimensional data based on Comparative DR [39]	IM, KPCA, LLE, Maximum Variance Unfolded	Protein Localisation Sites (E-Coli), Iris Dataset, Machine CPU Data, Thyroid Data
40	Linear DR for Multi-Label Classification [40]	PCA, LDA, CCA, Partial Least Squares(PLS) with SVM	Arts & Business Dataset
41	Research & Implementation of SVD [41]	SVD	Latent Semantic Indexing
42	A Survey on DR Techniques for Classification of Multi-dimensional data [42]	PCA, ICA, Factor Analysis, Non-Linear PCA< Random Projection, Auto Associative Neural networks	DR, Classification
43	Interpretable Dimension Reduction [43]	PCA	Cars Data
44	Deep Level Understanding of LDA [44]	LDA	Wine Data of Italy
45	Survey on ICA [45]	ICA	Statistics, Data Analysis, signal Processing
46	Image Reduction using Assorted DR Techniques [46]	PCA, Random Projection, LSA Transform, Many modified approaches	Images

for carrying out the research work have been listed out, along with the tools and techniques used in them. The table also includes the application areas covered by the respective papers.

8.3 Dimensionality Reduction Techniques

This section presents a detailed discussion over some of the most widely used algorithms for Dimension Reduction, which include both linear, and non-linear methods.

8.3.1 Principal Component Analysis

Principal Component Analysis (PCA) is a conventional unsupervised dimensionality reduction technique. With its wide range of applications, it has singlehandedly ruled over this domain for many decades. It makes use of Eigenvectors, Eigenvalues and the concept of variance of the data. Given a set of input variables, PCA aims at finding a new set of 'Y' variables:

$$y_i = f(X_i) = AX_i. \tag{8.1}$$

where A is the projection matrix, and dimn [Y] << dimn [X], such that a maximum portion of the information contained in the original set can be projected on this new set. For this, PCA computes unit orthonormal vectors, called Principal Components, which account for most of the variance of the data. The input data is observed as a linear combination of the principal components. These PCs serve as axes and thus, PCA can be defined as a method of creating a new coordinate system with axes $w_n \in$ RD (input space), chosen in a manner that the variance of the data is maximal, and:

$$w_n = arg_{||w||=1} \max var(Xw) = arg_{||w||=1} \max w' Cw. \tag{8.2}$$

For n=1,., i, the components can be calculated in the same manner. Here, X ∈ RD*N, is an input dataset of N samples and D variables, and C∈ RD*D is the covariance matrix of data X.

PCA can also be written as:

$$\max_{\{y\}} \sum_{i=1}^{n} || y_i - y_j ||^2 \ \ s.t. \ \ y_i = Ax_i \ \ and \ \ AA^T = I. \tag{8.3}$$

It is performed by conducting a series of elementary steps, which are:

(i) Firstly, normalisation of data points is done in order to create a standardised range of the variables. This is done by mean centering, i.e., subtracting the average value of each variable from it. This generates a zero mean data, i.e.,

$$\frac{1}{N}\sum_{i=1}^{n} x_i = 0. \tag{8.4}$$

where xi is the vector of one of the N multivariate observations. This step is necessary to avoid the probable chances of dominance of variables with large range over those with a comparably smaller range.

(ii) It is followed by creation of Covariance matrix. It is a symmetric matrix of the initial variables, of the order, n*n, where n=initial variables and:

$$C = \frac{1}{N}\sum_{i=1}^{n} x_i x_i^T. \tag{8.5}$$

It basically identifies the degree of correlation between the variables. The instances of this matrix are called variances. The Eigenvectors and eigenvalues of this matrix are computed, which further determine the Principal Components. These components are uncorrelated combinations of variables. The maximum information of the initial variables is contained in the first Principal Component, and then most of the remaining information is stored in the second component and this goes on.

(iii) Now, we choose the appropriate components and generating the feature vectors. The Principal Components are sorted in descending order on the basis of amount of variance carried by them. Now, the weaker components, the one with very low variance, are eliminated. The left components are used to build up a new dataset with reduced dimensionality. Generally, most of the variance is stored in first three or four components [14]. These components are then used to form the feature matrix. the percentage of variance accounted for by retaining the first q components is given by:

$$\frac{\sum_{k=1}^{q} \lambda_k}{\sum_{k=1}^{p} \lambda_k} \times 100. \tag{8.6}$$

Here, p refers to total initial eigenvalues, and λ_k is the variance of the kth instance.

Figure 8.3 shows a rough percent division of the variance of the data among the Principal Components. (This figure has been taken from an unknown online source.)

(iv) The last step involves re-casting of the data from original axes to the ones denoted by the Principal Components. It is simply done by performing multiplication of the transpose of the original dataset to the transpose of the feature vector.

The easy computational steps made it popular ever since 1930s, when it was developed. Due to its pliancy, it gathered a huge market within years of being released to the world. Its ability to handle large and multi-dimensional datasets is good, when compared to others at the same level.

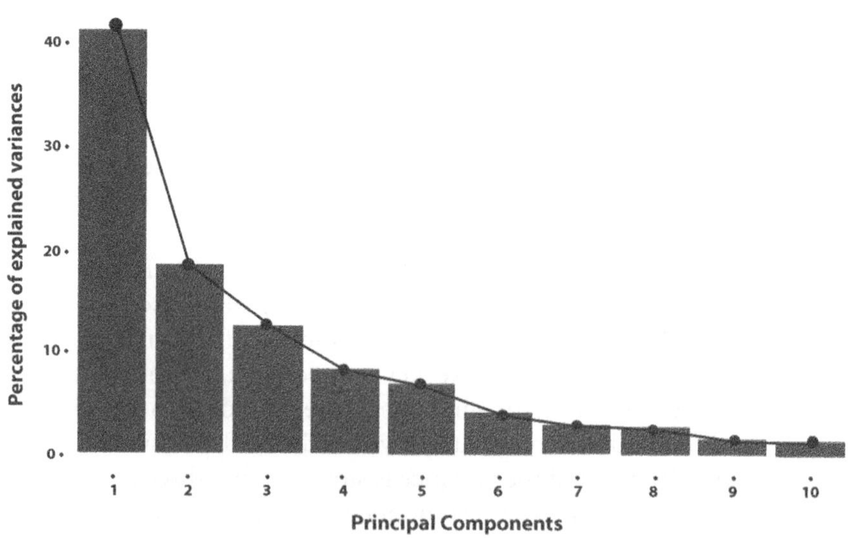

Figure 8.3 Five variances acquired by PCs.

Its application areas include signal processing, multivariate quality control, meteorological science, structural dynamics, time series prediction, pattern recognition, visualisation, etc. [11]. But it possesses certain drawbacks which hinder the expected performance. The linear nature of PCA provides unsatisfactory results with high inaccuracy when applied on non-linear data, and the fact that real world data is majorly non-linear, and complex worsens the situation. Moreover, as only first 2-3 Components are used generate the new variables, some information is always lost, which results in a not-so-good representation of data. Accuracy is affected due to this loss of information. Also, the size of covariance matrix increases with the dimensions of data points, which makes it infeasible to calculate eigenvalues for high dimensional data. To repress this issue, the covariance matrix can be replaced with the Euclidean distances. The Principal Components being a linear combination of all the input variables also serves as a limitation. The required computational time and memory is also high for PCA. Even after accounting for such drawbacks, it has given some fruitful results which cannot be denied. Soumya Raychaudhuri *et al.* [15] proved with a series of experiments that PCA was successful in finding reduced datasets when applied on sporulation datasets with better results, and also that it successfully identified periodic patterns in time series data.

These limitations can be overcome by bringing slight changes to the method. Some generalised forms of PCA have been created which vanquish its disadvantages, such as Sparse PCA, KPCA or Non-Linear PCA, Probabilistic PCA, Robust PCA, to name a few. Sparse PCA overcomes the disadvantage of PCs being a combination of all the input variables by adding a sparsity constraint on the input variables. Thus, making PCs a combination of only a few input variables. The Non-Linear PCA works on the nature of this traditional method and uses a kernel trick to make it suitable for non-linear datasets as well. Probabilistic PCA makes the method more efficient by making use of Gaussian noise model and a Gaussian prior. Robust PCA works well with corrupted datasets.

8.3.2 Linear Discriminant Analysis

Linear Discriminant Analysis (LDA), also known as discriminant function analysis, is one of the most commonly used linear dimensionality reduction techniques. It performs supervised dimensionality reduction by projecting input data to a linear subspace consisting of directions that maximise the separation between classes. In short, it produces a combination of variables or features in a linear manner, for characteristics of classes. Although, it should be duly noted that to perform LDA, continuous independent variables must be present, as it does not work on categorical independent variables.

LDA is similar to PCA but is supervised, PCA doesn't take labels into consideration and thus, is unsupervised. Also, PCA focuses on feature classification, on the other hand, LDA carries out data classification. LDA also overcomes several disadvantages of Logistics Regression, another algorithm for linear classification which is works well for binary classification problems. LDA can handle multi-class classification problems with ease.

LDA concentrates on maximising the distance among known categories and it does by creating a new axis in the case of Two-Class LDA and multiple axes in the case of Multi-Class LDA in a way to maximise the separation between known categories. The new axis/axes are created according to the following criteria which are considered simultaneously.

8.3.2.1 Two-Class LDA

(i) Maximise the distance between means of both categories.
(ii) Minimise the variation (which LDA calls "scatter") within each category (refer Figure 8.4).

8.3.2.2 Three-Class LDA

In the case of Multi-Class LDA, the number of categories/classes are more than two and there is a slight difference from the process used in Two-Class LDA:

(i) We first find the point that is central to all of the data.
(ii) Then measure the distances between a point that is central in each category and the main central point.
(iii) Now maximise the distance between each category and central point while minimising the scatter in each category (refer Figure 8.5).

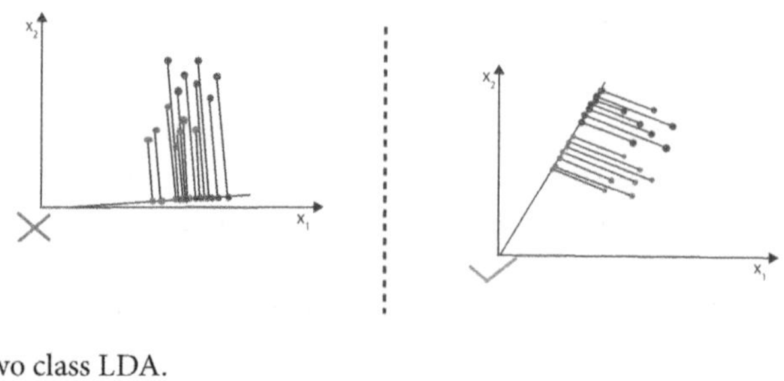

Figure 8.4 Two class LDA.

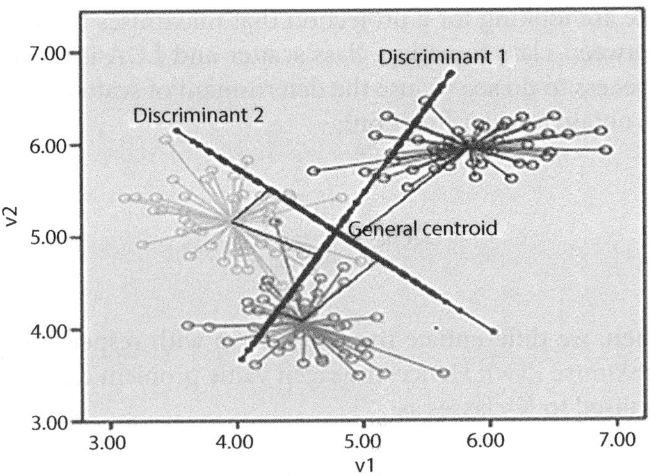

Figure 8.5 Choosing the best centroid for maximum separation among various categories.

While the ideas involved behind LDA are quite direct, but the mathematics involved is complex than those on which PCA is based upon. The goal is to find a transformation such that it will maximise the between-class distance and minimise the within-class distance. [Reference] For this we define two matrices: within-class scatter matrix and between-class scatter matrix.

The steps involved while performing LDA are:

(i) Given the samples X_1, X_2,……, X_n, and their respective labels y_1, y_2,……, y_n, the within-class matrix is computed as:

$$S_w = \sum_{i=1}^{n} (x_i - \mu_{yi})(x_i - \mu_{yi})^T. \tag{8.7}$$

where, $\mu_{yi} = \left(\frac{1}{N_i}\right)\sum_{x \in X_i}(x)$, (mean of y_i^{th} class) and N_i = number of data samples in class X_i.

(ii) The between-class matrix is computed as:

$$S_b = \sum_{k=1}^{m} n_k (x_i - \mu_{yi})(x_i - \mu_{yi})^T. \tag{8.8}$$

where $\mu = \frac{1}{N}\sum_{\forall X}(N_i\mu_i)$ (i.e. overall mean of whole sample), and $\mu_k = \frac{1}{N_i}\sum_{x \in X_i}(x)$ (i.e. mean of kth class).

(iii) We are looking for a projection that maximises the ratio of between-class to within-class scatter and LDA is actually a process to do so. We use the determinant of scatter matrices to obtain a scalar function:

$$Z(w)=\frac{\left|w^T S_b w\right|}{\left|w^T S_w w\right|}. \tag{8.9}$$

(iv) Then, we differentiate the above term with respect to w, to maximise Z(w). Hence, the eigen value problem can be generalised to K-classes as:

$$S_w^{-1} S_b w_i = \lambda w_i. \tag{8.10}$$

where, $\lambda_i = J(w_i)$ = scalar and i = 1, 2... (K-1).

(v) Finally, we sort the eigenvectors in a descending order and choose the top Eigenvectors to make our transformation matrix used to project our data.

This analysis is carried out by making a number of assumptions, which generates admirable results and leads to outperforming other linear methods. In [10], Paul Murray *et al.* showed how LDA was superior to PCA for performing experiments for inspection of rice-seed quality. These assumptions include multivariate normality, homogeneity of variance/covariance, multicollinearity and independence of participants' scores of features. LDA generates more accurate results when the sample sizes are equal.

The high applicability of LDA is a result of the advantages offered by it. Not only its ability to handle large and multi-class datasets is high, but also, it is less sensitive to faults. Also, it is very reliable when used on dichotomous features. It supports both binary and multi-class classifications. Apart from being the first algorithm used for bankruptcy prediction of firms, it has served as a pre-processing step in many applications such as statistics, bio-medical studies, marketing, pattern recognition, image recognition, and other machine learning applications. As any other technique, LDA also suffers from some drawbacks. While using LDA, lack of sample data leads to degraded classifier performance. A large number of assumptions in LDA also make it difficult for usage. Sometimes, it fails to preserve the complex structure of data, and is not suitable for non-linear mapping of data-points. LDA collapses when mean of the distributions are shared. This disadvantage came be eliminated by the use of Non-linear discriminant analysis.

Linear Discriminant Analysis has many extended forms, such as Quadratic Discriminant Analysis (QDA), Flexible Discriminant Analysis (FDA) and Regularised Discriminant Analysis (RDA). In Quadratic Discriminant Analysis, each class uses its own covariance/variance. In FDA, combinations of input are used in a non-linear Manner. RDA focusses on regularising the estimation of covariance.

8.3.3 Kernel Principal Component Analysis

Kernel Principal Component Analysis (KPCA) or Non-Linear PCA is one of the extended forms of PCA [11, 12]. The main idea behind this method is to modify the non-linear dataset in such a way that it becomes linearly separable. This is done by mapping the original dataset to a high-dimensional feature space, x→ ϕ(x) (ϕ is the non-linear mapping function of the sample x), and this results in a linearly separable dataset and now PCA can be applied on it for dimensionality reduction. But, carrying out calculations in the feature space can be very expensive, and basically infeasible, due to the high dimensionality of the feature space. So, we use kernel methods to carry out the mapping. It can be represented as follows:

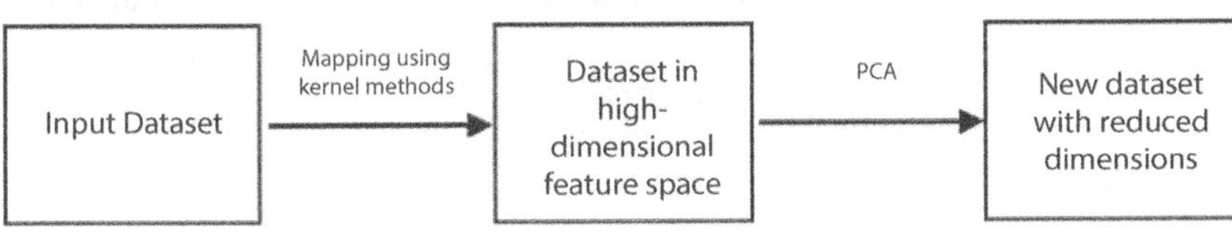

Figure 8.6 Kernel principal component analysis.

The kernel methods perform implicit mapping using a kernel function, which basically calculates the dot product of the feature vectors to perform non-linear mapping [11, 13] (refer Figure 8.6).

$$K(x_i, x_j) = \phi(x_i) \, . \, \phi(x_i)^T \tag{8.11}$$

Here, K is the kernel function. This is the central equation in the kernel methods. Now, the choice of kernel functions play a very significant roles the result of the entire method depends on it. Some of them are Linear kernel, Polynomial kernel, Sigmoid kernel, Radial Basis Function (RBF) kernel, Gaussian kernel, Spline kernel, Laplacian Kernel, Hyperbolic Tangent Kernel, Bessel kernel, etc. If the kernel function is linear, KPCA works similar to PCA and performs a linear transformation. When using polynomial kernel, the central equation can be stated as:

$$K(x_i, x_j) = (x_i \,.\, x_i + 1)^d. \tag{8.12}$$

Here, d is the degree of polynomial and we assume that the data points have zero mean. In Sigmoid kernel, which is popular in neural networks, the equation gets transformed to (with the assumption that the data points have zero mean):

$$K(x_i, x_j) = \tanh\left((x_i \,.\, x_i) + \theta\right). \tag{8.13}$$

The Gaussian kernel is used when there is no prior knowledge of the data. The equation used is:

$$K(x_i, x_j) = \exp\left(\frac{-\left(\left\|x_i - x_j\right\|\right)}{2\sigma^2}\right). \tag{8.14}$$

Here, again the data points are assumed to have zero mean. In case, the data points don't have zero mean, a normalisation content is added to the gaussian kernel's equation, $[\frac{1}{\sqrt{2\pi\sigma}}]^N$. Adding this constant makes the Gaussian kernel a normalised kernel, and the modified equation can be written as:

$$K(x_i, x_j) = [\frac{1}{\sqrt{2\pi\sigma}}]^N \exp\left(\frac{-\left(\left\|x_i - x_j\right\|\right)}{2\sigma^2}\right). \tag{8.15}$$

The Radial Basis Function (RBF) kernel is the most used due to its localised and finite response along the entire x-axis. It has many different types including Gaussian radial basis function kernel, Laplace radial basis function kernel, etc. the basic equation for the RBF kernel is:

$$K(x_i, x_j) = \exp(-\gamma \left\| x_i - x_j \right\|_2^2). \tag{8.16}$$

The procedure followed for execution of the KPCA method is:

(i) The initial step is to select a type of kernel function $K(x_i, x_j)$. ϕ is the transformation to higher dimension.

(ii) The Covariance matrix is generated after selecting the kernel function. In KPCA, the covariance matrix is called the Kernel matrix. It is generated by performing inner product of the mapped variables. It can one written as:

$$K = \phi(X) \,.\, \phi(X)^T. \tag{8.17}$$

This is called the kernel trick. It helps to avoid the necessity of explicit knowledge of ϕ.

(iii) The kernel matrix generated in the previous step is then normalised by using:

$$K' = K - 1_N K - K1_N + 1_N K1_N. \tag{8.18}$$

Here, 1N is a N*N matrix with all entries equal to (1/N). This step makes sure that the mapped features, using the kernel function, are zero-mean. Chis cantering operation performs subtraction of the mean of the data in feature space, defined by the kernel function.

(iv) Now, the eigenvectors and eigenvalues of the centred kernel matrix are calculated. The eigenvector equation is used to calculate and normalise the eigenvectors:

$$K' \alpha_i = \lambda_i \alpha_i. \tag{8.19}$$

Here, αi denotes the eigenvectors.

(v) The step is similar to the third step of the PCA. here, eigenvectors generate the Principal Components in the feature space, and further they are ranked in decreasing order on the basis of their eigenvalues. The Principal Component with the highest eigenvalue possesses maximum variance. Adequate components are then selected to map the data points on them in such a manner that the variance is maximised. The selected components are represented using a matrix.

(vi) The last step is to find the low dimensional representation which is done my mapping the data onto the selected components in the previous step. It can be done by finding the product of the initial dataset and the matrix obtained in the 5th step.

The results of de-noising images using linear PCA and KPCA have been shown in Figure 8.7. It can be observed that KPCA outperforms PCA in this case.

The kernel trick has been used in many techniques of the Machine Learning domain, such as Support vector machines, kernel ridge regression, etc. It has been proved useful for many applications, such as: Novelty detection, Speech recognition, Face recognition, Image de-noising, etc. The major advantage it offers is that it allows modification of linear methods to enable them to work on non-linear datasets and generate highly accurate results. Being a generalised version of PCA, KPCA owns all the advantages offered by PCA. Even though it overcomes the largest disadvantage of linear nature of PCA, it still has some limitations. To start with, the size of kernel mart is proportional to the square of variables of original dataset. On the top of this, KPCA focuses on retaining large pairwise distances. The training time required by this method is also very high. And due to its non-linear nature, it becomes more sensitive to fault when compared to PCA. Minh Hoai Nguyen *et al.* [17] proposed a robust extension of KPCA, called Robust KPCA, which showed better results for de-noising images, recovering missing data and handling intra-sample outliers. It outperformed other methods of same nature when experiments were conducted on various natural datasets. Many such methods have been proposed which mitigates the disadvantages offered by KPCA. Sparse KPCA is one of them. A. Lima *et al.* [13]

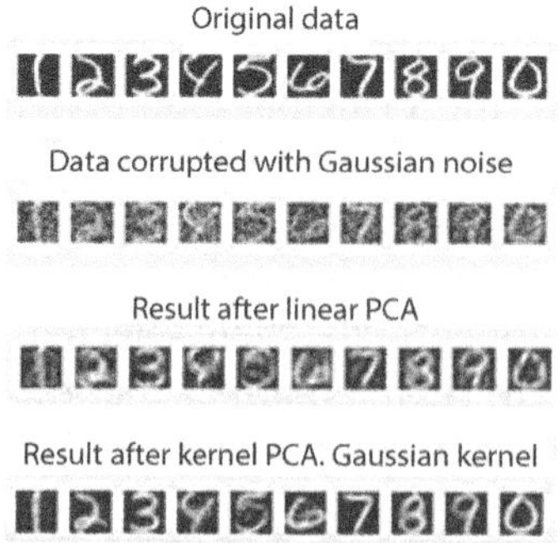

Figure 8.7 Results of de-noising handwritten digits.

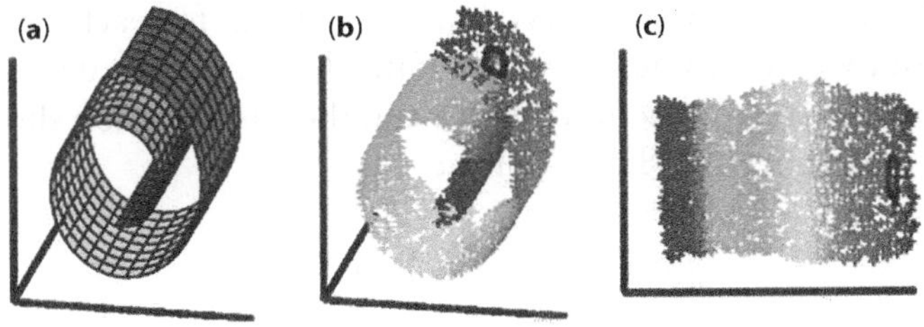

Figure 8.8 Casting the structure of Swiss Roll into lower dimensions.

proposed a version of Sparse KPCA for Feature Extraction in Speech Recognition. It treats the disadvantage of training data reduction in KPCA when the dataset is excessively large. This approach provided better results than PCA and KPCA on a Japanese ATR database (refer Figure 8.8).

8.3.4 Locally Linear Embedding

Locally Linear Embedding (LLE) is a non-linear technique for dimensionality reduction that preserves the local properties of data, it could mean preserving distances, angles or it could be something entirely different. It aims at maintaining the global construction of datasets by locally linear reconstructions. Being an unsupervised technique, class labels don't hold any importance for this analysis. Datasets are often represented in n-Dimensional feature space, with each dimension used for a specific feature. Many other algorithms of dimensionality reduction fail to be successful on non-linear space. LLE reduces these n-dimensions by preserving the geometry of the structure locally while piecing local properties together to preserve the structure globally. The resultant structure is casted into lower dimensions. In short, it makes use of local symmetries of the linear reconstructions to work with non-linear manifolds.

Simple geometric intuitions are the principle behind the working of LLE [Reference]. The procedure for Locally Linear Embedding algorithm includes three basic steps, which are as follows:

(i) LLE first computes the K nearest neighbours in which a point or a data vector is classified on basis of its nearest K neighbours but we have to careful while selecting the value of K, as K is the only parameter chosen and if too small or too big value is chosen, it will fail to preserve the geometry globally.

(ii) Then, a set of weights [W_{ij}] are computed, for each neighbour which denotes the effect of neighbour on that data vector. The weights cannot be zero and the cost function should be minimised as shown below:

$$E(W) = \Sigma_i |X_i - \Sigma_j W_{ij} X_j|^2. \tag{8.20}$$

Where jth is the index for nearest neighbour of point X_i.

(iii) Finally, we construct the low dimensional embedding of vector Y with the previously computed weights, and we do it by minimising the cost function below:

$$C(Y) = \Sigma_i |Y_i - \Sigma_i W_{ij} Y_j|^2. \tag{8.21}$$

In the achieved low- dimensional embedding, each point can still be represented with the same linear integration of its neighbours, as the one in the high dimensional representation.

LLE is an efficient algorithm particularly in pattern recognition tasks where the distance between the data points is an important factor in the algorithm and want to save computational time. LLE is widely used in pattern recognition, super-resolution, sound-source localisation, image processing problems and it shows significant results. It offers a number of advantages over other existing non-linear methods, such as: Non-linear PCA, Isomap, etc. Its ability to handle non-linear manifolds is commendable as it holds the capacity to identify a curved pattern in the structures of datasets. It even offers lesser computational time and memory as compared to other techniques. Also, it involves tuning only one parameter 'K' i.e., the number of nearest neighbours, therefore making the algorithm less complex. Although, some drawbacks of LLE exist, such as its poor performance when it encounters a manifold with holes. It also slumps large portions on data very close together when in the low dimensional representation. Such drawbacks have been removed by bringing slight modifications to the original analysis or generating extended versions of the algorithm. Hessian LLE (HLLE) is an example of an extension of LLE, which reduces the curviness of the original manifold while mapping it onto a low-dimensional subspace. Refer Figure 8.9 for Low dimensional Locally linear Embedding.

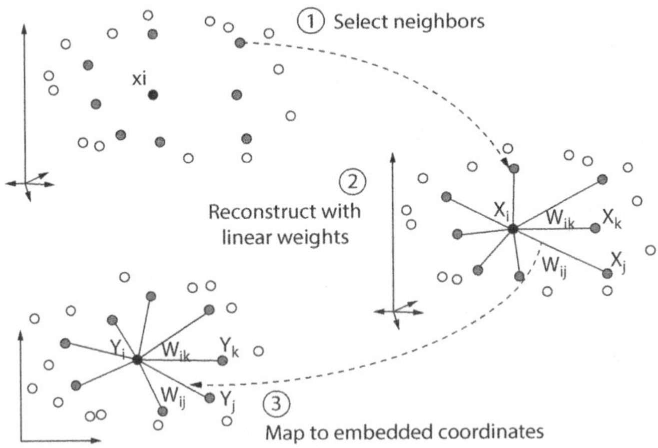

Figure 8.9 Working of LLE.

8.3.5 Independent Component Analysis

As we learned about PCA that it is about finding correlations by maximizing variances whereas in ICA we try to maximize independence by finding a linear transformation for our feature space into a new feature space such that each of the individual new features are mutually independent statistically. ICA does an excellent job in Blind Source Separation (BSS) wherein it receives a mixture of signals with very little information about the source signals and it separates the signals by finding a linear transformation on the mixture such that the output signals are statistically independent i.e. if sources{si} are statistically independent then:

$$p(s_1, s_2, .., s_n) = p(s_1)p(s_2), .., p(s_n). \tag{8.22}$$

Here, $\{s_i\}$ follows the non-gaussian distribution.

PCA does a poor job in Blind Source Separation. A common application of BSS is the cocktail party problem. The set of individual source signals are represented by $s(t) = \{s_1(t), s_2(t), ...s_n(t)\}$. Source signals (s(t)) are mixed with a mixing matrix (A) which produce the mixed signals (x(t)). So, mathematically we could express the relation as follows:

$$X(t) = \begin{vmatrix} x_1 \\ x_2 \end{vmatrix} = \begin{bmatrix} a & b \\ c & d \end{bmatrix} \begin{vmatrix} s_1 \\ s_2 \end{vmatrix} = A.s(t). \tag{8.23}$$

where, there are two signal sources (s_1 & s_2) and A (mixing matrix) contains the coefficients (a, b, c, d) of linear transformation.

The relation above is under some following assumptions:

- The mixing matrix (A) is invertible.
- The independent components have non-gaussian distributions.
- The sources are statistically independent.

To solve the above problem and recover our original strings from the mixed ones, we need to solve equation (1) for s(t) given by relation:

$$s(t) = A^{-1} \,.\, X(t). \tag{8.24}$$

Here, A^{-1} is called un-mixing matrix (W) and we need to find this inverse matrix to find our original sources and choose the numbers in this matrix in such a way that maximizes the probability of our data.

Independent Component Analysis is used in multiple fields and applications such as telecommunications, stock prediction, seismic monitoring, text document analysis, optical imaging of neurons and often applied to reduce noise in natural images.

8.3.6 Isometric Mapping (Isomap)

Isomap (IM), short for Isometric mapping, is a non-linear extended version of Multidimensional Scaling (MDS). It focuses on preserving the overall geometry of the input dataset, by making use of a weighted neighbourhood graph 'G' for performing low dimensional embedding of the initial data in high-dimensional manifold. Unlike MDS, it aims at sustaining the Geodesic pairwise distance between all the data points. The concept and procedure followed by Isomap is very similar to Locally Linear Embedding (LLE), except the fact that the latter focuses on maintaining the local structure of the data while carrying out transformation, whereas Isomap is more inclined towards conserving the global structure along with the local geometry of the data points.

The IM algorithm executes the following three steps to procure a low dimensional embedding:

(i) The procedure starts with the formation of a neighbourhood weighted graph G, by considering 'k' nearest neighbours of the data points xi (i=1, 2,...,n), where the edge weights are equal to the Euclidean distances. This steps ensures that local structure of the dataset does not get compromised.

(ii) The next step is to determine the geodesic distances, and form a Geodesic distance matrix. Geodesic distance can be defined as the sum of edge weights and the shortest path between two data points. This is done by making use of Dijkstra's algorithm or Floyd-Warshall shortest path algorithm. It is the distinguishing step between Isomap and MDS.

(iii) The last step is to apply MDS on the matrix obtained in the previous step.

Preserving the curvilinear distances over a manifold is the biggest advantage offered by Isomap as usage of Euclidean distances over a curved manifold can generate misleading results. Geodesic distance helps to overcome this issue faced by MDS. Isomap has been successfully applied to various applications such as: Pattern Recognition, Wood inspection, Image processing, etc. A major flaw Isomap suffers with is short circuiting errors, which occur due to inaccurate connectivity in the graph G. A. Saxena *et al.* [28] overcame this issue by removing certain neighbours that caused issues in determining the local linearity of the graph. It has also failed under circumstances where the manifold was non-convex and if it contains holes.

Many Isomap generalisations have been created over the years, which include: Conformal Isomap, Landmark Isomap and Parallel transport unfolding. Conformal Isomap or C-Isomap owns the ability to understand curved manifold in a better way, by magnifying highly dense sections of the manifold, and narrowing down the regions with less intensity of data points. Landmark Isomap (L-Isomap) reduces the computational complexity by considering a marginal amount of landmark points out of the entire set. Parallel transport unfolding works on removing the voids and irregularity in sampling by substituting the geodesic distances for parallel transport-based approximations. In [8], Vin de Silva *et al.* presented an improved approach to Isomap and derived C-Isomap and L-Isomap algorithms which exploited computational sparsity.

8.3.7 Self-Organising Maps

Self-Organising Map (SOM) are unsupervised neural networks that are used to project high-dimensional data into low-dimensional output which is easy to visualize and understand. Ideas were first introduced by C. von der Malsburg in 1973 but developed and refined by T. Kohonen in 1982. SOMs are mainly used for clustering (or classification), data visualization, probability, modelling and density estimation. There are no hidden layers in these neural networks and only contains an input and output layer.

SOM uses Euclidean distances to plot data points and the neutrons are arranged on 2-dimensional grid also called as a map.

First, we initialize neural network weights randomly and choose a random input vector from training dataset and also set a learning rate (η).

Then for each neuron j, compute the Euclidean distance:

$$D(j)=\sqrt{\Sigma n(x_i-w_{ij})^2}. \tag{8.25}$$

Here, x_i is the current input vector and w_{ij} is the current weight vector.

We then select the winning neutron (Best Matching Unit) with index j such that D(j) is minimum and then we update the network weights given by the equation:

$$W_{ij}(new) = W_{ij}(old) + \theta_{ij}(t)\eta(t)(X_i - W_{ij}(old)). \tag{8.26}$$

Here, (t) (learning rate) = $\eta_0$exp $(-\ t\ /\lambda)$, t = epoch, λ = time constant. The learning rate decay is calculated for every epoch.

$$\theta_{ij}(t)(influence\ rate)=\exp\left(\frac{-D(j)2}{2\sigma 2(t)}\right). \tag{8.27}$$

Where, σ is called the Neighbourhood Size which keeps on decreasing as the training continues given by an exponential decay function:

$$\sigma(t)=\sigma_0\ exp(\frac{-t}{\lambda}). \tag{8.28}$$

The influence rate signifies the effect of a node distance from the selected neutron (BMU) has its own learning and finally through many iterations and updating of weights, SOM reaches a stable configuration.

Self-organising maps are applied to wide range of fields and applications such as in analysis of financial stability, failure mode and effect analysis, classifying world poverty, seismic facies analysis for oil and gas exploration etc. and is a very powerful tool to visualize the multi-dimensional data.

8.3.8 Singular Value Decomposition

SVD is a linear dimensionality reduction technique which basically gives us the best axis to project our data in which the sum of squares of projection error is minimum. In other words, we can say that it allows us to rotate the axes in which the data is plotted to a new axis into a new axis along the directions that have maximum variance. It is based on simple linear algebra which makes it very convenient to use it on any data matrix where we have to discover latent, hidden features and any other useful insights that could help us in classification or clustering. In SVD an input data matrix is decomposed into three unique matrices:

$$A_{[m\times n]} = U_{[m\times n]} \Sigma_{[m\times n]}(V_{[m\times n]})^T. \tag{29}$$

where A: [m x n] input data matrix, U: [m x m] real or complex unitary matrix (also called left singular vectors), Σ: [m x n] diagonal matrix, and V: [n x n] real or complex unitary matrix (also called right singular vectors).

U and V are column orthonormal matrices, meaning the length of each column vector is one. The values in Σ matrix is called singular values and they are positive and sorted in decreasing order, meaning the largest singular values come first.

SVD is widely used in many different applications like in recommender systems, signal processing, data analysis, latent semantic indexing and pattern recognition etc. and is also used in performing Principal Component Analysis (PCA) in order to find principal directions which, have the maximum variance. Also, the rotation in SVD helps in removing collinearity in the original feature space. SVD doesn't always work well specially in cases of strongly non-linear data and its results are not ideal for good visualizations and while it is easy to implement the algorithm but at the same time it is computationally expensive.

8.3.9 Factor Analysis

Factor Analysis is a variable reduction technique which primarily aims at removing highly redundant data in our dataset. It does so by removing highly correlated variable into small numbers of latent factors. Latent factors are the factors which are not observed by us but can be deduced from other factors or variables which are directly observed by us. There are two types of Factor Analysis: Exploratory Factor Analysis and Confirmatory Factor Analysis. The former focuses on exploring the pattern among the variables with no prior knowledge to start with while the later one is used for confirming the model specification.

Consider the following matrix equation from which Factor analysis assumes its observable data that has been deduced from latent factors:

$$y = (x - \mu) = LF + \varepsilon. \tag{8.30}$$

Here, x is a set of observable random variables with means μ. L contains the unknown constants and F contains "Common Factors" which are unobserved random variables and influences the observed variables. ε is the unobserved error terms or the noise which is stochastic and have a finite variance.

The common factors matrix(F) is under some assumptions:

- F and ε are independent.
- Corr(F) = I (Identity Matrix), here, "Corr" is the cross-covariance matrix.
- E(F) = 0 (E is the Expectation).

Under these assumptions, the covariance matrix of observed variables [Reference] is:

$$Corr(y) = LCorr(F)LT + Corr(\varepsilon). \tag{8.31}$$

Taking Corr(y) = Σ and Corr(ε) = λ, we get Σ = LLT + λ. The matrix L is solved by the factorization of matrix LLT = Σ - λ.

We should consider that prior to performing Factor Analysis the variables are following multivariate normal distribution and there must be large number of observations and enough number of variables that are related to each other in order to perform data exploration to simplify the given dataset but if observed variables are not related, factor analysis will not be able to find a meaningful pattern among the data and will not be useful in that case. Also, the factors are sometimes hard to interpret so it depends on researcher's ability to understand it attributes correctly.

8.3.10 Auto-Encoders

Auto-Encoders are unsupervised practical implementation of otherwise supervised neural networks. Neural networks are basically a string of algorithms, that try to implement the way in which human brain processes the gigantic amount of data. In short, neural networks tend to identify the underlying pattern behind how the data is related, and thus perform classification and clustering in a way similar to a human brain. Auto-encoder performs dimensionality reduction by achieving reduced representation of the dataset with the help of a bottleneck, also called the hidden layer(s).

The first half portion of an auto-encoder encodes the data to obtain a compressed representation, while the second half focuses on regenerating the data from the encoded representatives. The simplest form of an auto-encoder consists of three layers: The Input layer, the hidden layer (bottleneck) and the output layer.

The architecture of an auto-encoder can be well explained in two steps:

(i) Encoder: This part of an auto-encoder accepts the input data, using the input layer. Let x ∈ Rd be the input. The hidden layer (bottleneck) maps this data onto H, such that H ∈ RD. where H is the low dimensional representation of the input X. Also,

$$H = \rho(Wx + b). \tag{8.32}$$

Where ρ is the activation function, W denotes the Weight matrix and b is the bias vector.

(ii) Decoder: This part is used for reconstruction of the data from the reduced formation achieved in the previous step. The output generated by it is expected to be the same as the input. Let x′ be the reconstruction, which is of the same shape as x, then x′ can be represented as:

$$x' = \rho'(W'H + b'). \tag{8.33}$$

Here, ρ′, W′ and b′ might not be same as in equation (8.32).

The entire auto-encoder working can be expressed in the following equations:

$$\phi: X \rightarrow F. \tag{8.34}$$

$$\psi: F \rightarrow X'. \tag{8.35}$$

$$\phi, \psi = arg\ min||X - (\psi.\phi)X||^2. \tag{8.36}$$

Where F is the feature space and H ∈ F, ϕ and ψ are the transitions in the two phases and X and X' are the input and output spaces, which are expected to coincide perfectly.

The existence of more than 1 hidden-layers give birth to Multilayer auto-encoders. The concept of Auto-encoders has been successfully applied various applications which include information retrieval, image processing, Anomaly detection, HIV analysis etc. It makes use of the phenomenon of back-propagation to minimise the reconstruction loss, and also for training of the auto-encoder. Although back propagation converges with increasing number of connections, which serves as a drawback. It is overcome by pre-training of the auto-encoder, using RBMs. In [9], Omprakash Saini *et al.* stated poor interpretability as one of its other drawbacks, and pointed out various other advantages, such as, its ability to adopt parallelization techniques for improving the computations.

8.4 Experimental Analysis

8.4.1 Datasets Used

In following experiments, we reduce the feature set of two different datasets using both linear and non-linear dimension reduction techniques. We would also compute accuracy of predictions of each technique and lastly compare the performance of techniques used in this experimental analysis.

Datasets used are as following:

- Red-Wine Quality Dataset: The source of this dataset is UCI which is a Machine Learning repository. The wine quality dataset has two datasets, related to red and white wine samples of Portugal wines. For this paper, Red wine dataset issued which consists of 1599 instances and 12 attributes. It can be viewed as classification and regression tasks.
- Wisconsin Breast Cancer Dataset: This dataset was also taken from UCI, a Machine Learning repository. It is a multivariate dataset, containing 569 instances, 32 attributes and no missing values. The features of the dataset have been computed by using digitised images of FNA of a breast mass.

8.4.2 Techniques Used

- Linear Dimensionality Reduction Techniques: Principal Component Analysis (PCA), Linear Discriminant Analysis (LDA), Independent Component Analysis (ICA), Singular Value Decomposition (SVD).

- Non-Linear Dimensionality Reduction Techniques: Kernel Principal Component Analysis (KPCA), Locally Linear Embedding (LLE).

8.4.3 Classifiers Used

- In case of Red Wine Quality Dataset, Random Forest algorithm is used to predict the quality of red wine.
- For prediction in Wisconsin Breast Cancer Dataset, Support-Vectors Machine (SVM) classifier is used.

8.4.4 Observations

Dimensionality Reduction Techniques Results on RED-WINE Quality Dataset (1599 rows X 12 columns), using Random Forest as classifier, have been shown in Table 8.2.

Table 8.3 shows the Dimensionality Reduction Techniques Results on WISCONSIN BREAST-CANCER Quality Dataset (569 rows X 33 columns) using SVM as classifier.

8.4.5 Results Analysis Red-Wine Quality Dataset

- Both PCA and LDA shows the highest accuracy of 64.6% correct predictions among all the techniques used.
- Both the techniques reduce the dimensions of dataset from 12 to 3 most important features.
- Non-Linear techniques used i.e. KPCA & LLE doesn't perform well on this dataset and all the Linear Dimensionality Reductions techniques outperformed the non-linear techniques.

Wisconsin Breast Cancer quality dataset

- PCA technique shows the best accuracy among all the techniques with an error rate of only 2.93%, which means over 97% of the cases were predicted correctly.
- PCA reduces the dimension of dataset from 33 features to 5 most important features to achieve its accuracy.
- Again, the Linear Reduction techniques outperformed the non-linear techniques used in this dataset.

Table 8.2 Results of red-wine quality dataset.

Dimension reduction techniques	Total number of data rows	Number of actual dimensions	Number of reduced dimensions	Correct prediction %	Error %
PCA	1599	12	3	64.6%	35.4%
LDA	1599	12	3	64.6%	35.4%
KPCA	1599	12	1	44.06%	55.94%
LLE	1599	12	1	42.18%	57.82%
ICA	1599	12	3	65.31%	34.69%
SVD	1599	12	3	64.48%	35.52%

Table 8.3 Results of Wisconsin breast cancer quality dataset.

Dimension reduction techniques	Total number of data rows	Number of actual dimensions	Number of reduced dimensions	Correct prediction %	Error %
PCA	569	33	5	97.07%	2.93%
LDA	569	33	3	95.9%	4.1%
KPCA	569	33	1	87.71%	12.29%
LLE	569	33	1	87.13%	12.87%
ICA	569	33	3	70.76%	29.24%
SVD	569	33	4	95.9%	4.1%

8.5 Conclusion

Although, researchers have been working on finding techniques to cope up with the high dimensionality of data, which serves as a disadvantage, for more than a hundred years now, the challenging nature of this task has evolved with all the progress in this field. Researchers have come a long way since 1900s, when the concept of PCA first came into existence. However, from the experiments performed for this research work, it can be concluded that the linear and the traditional techniques of Dimensionality Reduction still outperform the non-linear ones. This conclusion is apt for most of the datasets. The results generated by PCA make it the most desirable tool. The error percentage of the contemporary, non-linear techniques make them inapposite. Having said that, research work is still in its initial stages for the huge, non-linear datasets and proper exploration and implementation of these techniques can lead to generation of fruitful results. In short, the benefits being offered by the non-linear techniques can be fully enjoyed by doing more research and improving the pitfalls.

References

1. Mishra, P.R. and Sajja, D.P., Experimental survey of various dimensionality reduction techniques. *Int. J. Pure Appl. Math.*, 119, 12, 12569–12574, 2018.
2. Sarveniazi, A., An actual survey of dimensionality reduction. *Am. J. Comput. Math.*, 4, 55–72, 2014.
3. Bartenhagen, C., Klein, H.-U., Ruckert, C., Jiang, X., Dugas, M., Comparative study of unsupervised dimension reduction techniques for the visualization of microarray gene expression data. *BMC Bioinf.*, 11, 1, 567–577, 2010.
4. Patil, M.D. and Sane, S.S., Dimension reduction: A review. *Int. J. Comput. Appl.*, 92, 16, 23–29, 2014.
5. Globerson, A. and Tishby, N., Most informative dimension reduction. *AAAI-02: AAAI-02 Proceedings*, pp. 1024–1029. Edmonton, Alberta, Israel, August 1, 2002.
6. Nelson, D. and Noorbaloochi, S., Sufficient dimension reduction summaries. *J. Multivar. Anal.*, 115, 347–358, 2013.
7. Ma, Y. and Zhu, L., A review on dimension reduction. *Int. Stat. Rev.*, 81, 1, 134–150, 2013.
8. de Silva, V. and Tenenbaum, J.B., Global versus local methods in nonlinear dimensionality reduction. *NIPS'02: Proceedings of the 15th International Conference on Neural Information Processing*, pp. 721–728, MIT Press, MA, United States, 2002.

9. Saini, O. and Sharma, P.S., A review on dimension reduction techniques in data mining. *IISTE*, 9, 1, 7–14, 2018.
10. Fabiyi, S.D., Vu, H., Tachtatzis, C., Murray, P., Harle, D., Dao, T.-K., Andonovic, I., Ren, J., Marshall, S., Comparative study of PCA and LDA for rice seeds quality inspection. *IEEE Africon*, pp. 1–4, Accra, Ghana, *IEEE*, September 25, 2019.
11. Tippin, M.E., Sparse kernel principal component analysis. *NIPS'00: Proceedings of the 13th International Conference on Neural Information Processing Systems*, United States, MA, January 2000, MIT Press, pp. 612–618,, MIT Press, United States, MA, January 2000.
12. Kim, K., II, Jung, K., Kim, H.J., Face recognition using kernel principal component analysis. *IEEE Signal Process. Lett.*, 9, 2, 40–42, 2002.
13. Lima, A., Zen, H., Nankaku, Y., Tokuda, K., Kitamura, T., Resende, F.G., Sparse KPCA for feature extraction in speech recognition. *IEICE Trans. Inf. Syst.*, 1, 3, 353–356, 2005.
14. Yeung, K.Y. and Ruzzo, W.L., Principal component analysis for clustering gene expression data. *OUP*, 17, 9, 763–774, 2001.
15. Raychaudhuri, S., Stuart, J.M., Altman, R.B., Principal component analysis to summarize microarray experiments: Application to sporulation time series. *Pacific Symposium on Biocomputing*, vol. 5, pp. 452–463, 2000.
16. Hinton, G.E. and Salakhutdinov, R.R., Reducing the dimensionality of data with neural networks. *Sci. AAAS*, 313, 5786, 504–507, 2006.
17. Nguyen, M.H. and De la Torre, F., Robust kernel principal component analysis, in: *Advances in Neural Information Processing Systems 21: Proceedings of the Twenty-Second Annual Conference on Neural Information Processing Systems*; Vancouver, British Columbia, Canada, December 8-11, 2008. 185-119, Curran Associates, Inc., NY, USA, 2008.
18. Raymer, M.L., Punch, W.F., Goodman, E.D., Kuhn, L.A., Jain, A.K., Dimensionality reduction using genetic algorithms. *IEEE Trans. Evol. Comput.*, 4, 2, 164–171, 2000.
19. DeMers, D. and Cottre, G., Non-linear dimensionality reduction. *Advances in Neural Information Processing Systems,* 5, 1993; 580-587, NIPS, Denver, Colorado, USA, 1992.
20. Tenenbaum, J.B., de Silva, V., Langford, J.C., A global geometric framework for nonlinear dimensionality reduction. *Sci. AAAS*, 290, 5500, 2319–2323, 2000.
21. Zhang, D., Zhou, Z.-H., Chen, S., Semi-supervised dimensionality reduction. *Proceedings of the Seventh SIAM International Conference on Data Mining*, Minneapolis, Minnesota, USA, April 26-28, Society for Industrial and Applied Mathematics, 3600 University City Science Center, Philadelphia, PA, United States, pp. 11–393, 2007.
22. Sarwar, B.M., Karypis, G., Konstan, J.A., Riedl, J.T., Application of dimensionality reduction in recommender system–a case study. *ACM WEBKDD*

Workshop: Proceedings of ACM WEBKDD Workshop, USA, 2000, p. 12, Association for Computing Machinery, NY, USA, 2000.

23. Raich, R., Costa, J.A., Damelin, S.B., Hero III, A.O., Classification constrained dimensionality reduction. *ICASSP: Proceedings ICASSP 2005*, Philadelphia, PA, USA, March 23, 2005, IEEE, NY, USA, 2005.
24. van der Maaten, L.J.P., Postma, E.O., van den Herik, H.J., Dimensionality reduction: A comparative review. *J. Mach. Learn. Res.*, 10, 1, 24, 66–71, 2007.
25. Adragni, K.P. and Cook, R.D., Sufficient dimension reduction and prediction in regression. *Phil. Trans. R. Soc. A*, 397, 4385–4405, 2009.
26. Alam, M.A. and Fukumizu, K., Hyperparameter selection in kernel principal component analysis. *J. Comput. Sci.*, 10, 7, 1139–1150, 2014.
27. Wang, Q., Kernel principal component analysis and its applications in face recognition and active shape models. *Corr*, 1207, 3538, 27, 1–8, 2012.
28. Hamdi, N., Auhmani, K., M'rabet Hassani, M., Validation study of dimensionality reduction impact on breast cancer classification. *Int. J. Comput. Sci. Inf. Technol.*, 7, 5, 75–84, 2015.
29. Vlachos, M., *Dimensionality reduction KDD '02: Proceedings of the eighth ACM SIGKDD International Conference on Knowledge Discovery and Data Mining* Edmonton, Alberta, Alberta 2002, pp. 645–651, Association for Computing Machinery, NY, United States, 2002.
30. Sembiring, R.W., Zain, J.M., Embong, A., Dimension reduction of health data clustering. *Int. J. New Comput. Archit. Appl.*, 1, 3, 1041–1050, 2011.
31. Wang, W. and Carreira-Perpinan, M.A., The role of dimensionality reduction in classification. *AAAI Conference on Artificial Intelligence*, Québec City, Québec, Canada, July 27–31, 2014, AAAI Press, Palo Alto, California, pp. 1–15, 2014.
32. Cunningham, P., Dimension reduction, in: *Technical Report UCD-CSI*, pp. 1–24, 2007.
33. Partridge, M. and Sedal, R.C., Fast dimensionality reduction and simple PCA. *Intell. Data Anal.*, 2, 3, 203–214, 1998.
34. Voruganti, S., Ramyakrishna, K., Bodla, S., Umakanth, E., Comparative analysis of dimensionality reduction techniques for machine learning. *Int. J. Sci. Res. Sci. Technol.*, 4, 8, 364–369, 2018.
35. Varghese, N., Verghese, V., Gayathri, P., Jaisankar, D.N., A survey of dimensionality reduction and classification methods. *IJCSES*, 3, 3, 45–54, 2012.
36. Fodor, I.K., *A Survey of Dimension Reduction Techniques*, pp. 1–18, Center for Applied Scientific Computing, Lawrence Livermore National Laboratory, 2002.
37. Roweis, S.T. and Saul, L.K., Nonlinear dimensionality reduction by locally linear embedding. *Sci. AAAS*, 290, 5500, 2323–2326, 2000.
38. Govinda, K. and Thomas, K., Survey on feature selection and dimensionality reduction techniques. *Int. Res. J. Eng. Technol.*, 3, 7, 14–18, 2016.
39. Sembiring, R.W., Zain, J.M., Embong, A., Alternative model for extracting multidimensional data based-on comparative dimension reduction, in:

CCIS: Proceedings of International Conference on Software Engineering and Computer Systems, Pahang, Malaysia, June 27-29, 2011 Springer, Berlin, Heidelberg, Malaysia, pp. 28–42, 2011.

40. Ji, S. and Ye, J., Linear dimensionality reduction for multi-label classification, in: *Twenty-First International Joint Conference on Artificial Intelligence*, Pasadena, California, June 26, 2009, AAAI Press, pp. 1077–1082, 2009.
41. Wang, Y. and Lig, Z., Research and implementation of SVD in machine learning. *IEEE/ACIS 16th International Conference on Computer and Information Science (ICIS)*, Wuhan, China, May 24-26, 2017, IEEE, NY, USA, pp. 471–475, 2017.
42. Kaur, S. and Ghosh, S.M., A survey on dimension reduction techniques for classification of multidimensional data. *Int. J. Sci. Technol. Eng.*, 2, 12, 31–37, 2016.
43. Chipman, H.A. and Gu, H., Interpretable dimension reduction. *J. Appl. Stat.*, 32, 9, 969–987, 2005.
44. Ali, A. and Amin, M.Z., A Deep Level Understanding of Linear Discriminant Analysis (LDA) with Practical Implementation in Scikit Learn, pp. 1–12, Wavy AI Research Foundation, 2019. https://www.academia.edu/41161916/A_Deep_Level_Understanding_of_Linear_Discriminant_Analysis_LDA_with_Practical_Implementation_in_Scikit_Learn
45. Hyvarinen, A., Survey on independent component analysis. *Neural Comput. Surv.*, 2, 4, 94–128, 1999.
46. Nsang, A., Bello, A.M., Shamsudeen, H., Image reduction using assorted dimensionality reduction techniques. *Proceedings of the 26th Modern AI and Cognitive Science Conference*, Greensboro, North Carolina, USA, April 25–26, 2015, MAICS, Cincinnati, OH, pp. 121–128, 2015.

9

Big Data Analytics in Real Time for Enterprise Applications to Produce Useful Intelligence

Prashant Vats[1] and Siddhartha Sankar Biswas[2]*

[1]*Department of Computer Science & Engineering, Faculty of Engineering & Technology, SGT University, Gurugram, Haryana, India*
[2]*Department of Computer Science & Engineering, Jamia Hamdard, New Delhi, India*

Abstract

Big data is a technique for storing and analyzing massive amounts of data. The use of this technical edge allows businesses and scientists to focus on revolutionary change. The extraordinary efficacy of this technology outperforms database management systems based on relational databases (RDBMS) and provides a number of computational approaches to help with storage bottlenecks, noise detection, and heterogeneous datasets, among other things. It also covers a range of analytic and computational approaches for extracting meaningful insights from massive amounts of data generated from a variety of sources. The ERP or SAP in data processing is a framework for coordinating essential operations and with the customer relationship and supply chain management. The business arrangements are transferred to optimize the whole inventory network. Despite the fact that an organization may have a variety of business processes, this article focuses on two continuous business use cases. The first is a data-processing model produced by a machine, the general design of this method, as well as the results of a variety of analytics scenarios. A commercial agreement based on diverse human-generated data is the second model. This model's data analytics describe the type of information needed for decision making in that industry. It also offers a variety of new viewpoints on big data analytics and computer techniques. The final section discusses the difficulties of dealing with enormous amounts of data.

Keywords: Big data, IoT, business intellectual, data integrity, industrial production

**Corresponding author*: ssbiswas@jamiahamdard.ac.in

M. Niranjanamurthy, Kavita Sheoran, Geetika Dhand, and Prabhjot Kaur (eds.) Data Wrangling: Concepts, Applications and Tools, (187–212) © 2023 Scrivener Publishing LLC

9.1 Introduction

In recent decades, IoT technology and data science have become the most discussed technologies on the planet. These two developments work together to collect data regularly. IoT will vastly increase the amount of information available for investigation by all sorts of organizations. Regardless, there are still significant issues to overcome even before anticipated advantages may be fully realized. The Internet of Things (IoT) and big data are certainly developing rapidly, and they are causing changes in a variety of industries and also in everyday situations. Due to the obvious connection of sensors, the Internet of Things generates a massive input of massive data. The Internet of Things will determine the course of business intelligence tools. Organizations can deliver memorable sector reforms by drawing actionable intelligence from massive amounts of data. The fundamental concept is to deploy IoT onto Business Applications in industrial automation. To identify the needs of something like the decisional analytical support system in the cloud environments, the manufacturing process requirement must be considered. The enterprise concept and present IT setup are investigated to identify methodological gaps in using the concept of the Internet of Things as a framework for smart manufacturing. IoT opens the way for industrial businesses to grow by enhancing existing systems in a globalized and fragmented environment. In any event, IoT operations are still in their early stages in many businesses, and more study is needed before moving forward with deployment.

The potential of big data is speculated by inside and then remotely gathered information using linked gadgets. The Internet of Things refers to Internet connections to the physical universe and ordinary things. This advancement brings up a plethora of new possibilities. Because embedding infrastructures and informational technologies are directly integrated into the transition, smart physical devices play an important role in the concept of IoT technology. IoT may be defined as the connecting of physical worlds, detectors within and connected with objects, and the internet via remotely through hardwired system connections. The phrase "data science" refers to the massive amounts of data administration and the provision of information via an inquiry that exceeds the capability of traditional database management systems (RDBMS). Data science is not only changing data storage and administration techniques but also provides competent continual analytics and graphic representations, which imply the qualitative information necessary for the enterprise. Big data is becoming increasingly essential for many companies. The activities necessitate a broader range of applications that can handle an increasing number of data that is constantly generated from different data sources. Data science manages data that cannot be used or handled in a typical manner. A conventional DBMS has less storage, it is harder to address problems in the data set, and processing is quite simple. However, in the case of huge data, special emphasis is required for data cleansing and the calculating method. Continuous data pouring necessitates decisions regarding which parts of the streaming data should be captured for analytical purposes.

Data science analytics should be lauded for freeing a competitive advantage over its competitor's market for the benefit of the company. In the conventional framework, interpretations from various findings, such as sales reports and inventory status may be captured using readily available business predictive analytical tools. The combination of conventional and big data decides the actionable, intelligent analytical findings required by the organization. Consequently, scheduling and forecasting apps derive information from big data. To generate understanding from this vast amount of big data, businesses must use data analytics. The word analytics is most commonly used to refer to data-driven decision making. The assessment is used for both business and academic research. Although those are separate types of research, the identical data contained in commercial examination necessitates knowledge in data mining, commercial factual methods, and visualization to satisfy the inquiries of corporate visionaries. Analytics plays an important role in gaining valuable understandings of business operations and finance. It should look into the requests made by consumers, items, sales, and so forth. The combination of corporate interests' information and big data aided in predicting the behavior of customers in the selection of the materials. In any event, whenever an instance of a scholarly article occurs, these must always be examined to investigate the hypothesis and create new ideas. Industrial revolution 4.0 is a contemporary transformation that is paving the way for IoT-based smart industrial production. Integrating IoT and data science is indeed a multidisciplinary activity that needs a special set of skills to provide the most extreme benefits from frameworks. Intellectual networking may be built up in the production process framework to link, manage, and correlate to one another automatically with significantly decreased interference by administrators. It also has the tangible potential to affect important company necessities and is now in the process of renovating industrial segments. Data wrangling analytics is a way of bursting large volumes of data that contain many types of information, i.e., big data, to expose all underlying patterns, undiscovered linkages, industry trends, customer decisions, as well as other useful enterprise data.

The results of said analytics can lead to much more clever advertising, new enterprise possibilities, and better customer service, as well as increased performance improvement, gain competitive advantage, and other economic advantages. The primary goal of predictive analytics is to assist organizations in making quite beneficial management decisions by enabling data researchers, analytics professionals, and other business intelligence experts to analyze large amounts of data from various operations, as well as other kinds of data that may go unnoticed by other more typical Business Intellectual capacity (BI) programs. Website logs, social networking sites, online trade, online communities, web click information, mails from consumers, survey results, mobile telephone call records, and machine data created by gadgets connected with IoT-based networks may all be included. This chapter describes the application of IoT, data science, and other analytical tools and methods for exploiting the massive volume of structured and unstructured data generated in the commercial setting. Data wrangling-based business intelligence plays an important role in achieving extraordinary results by offering cognitive insights from accessible data to utilize operations and business expertise. It provides accurate historical trends, as well as online monitors for effective decision making throughout the enterprise's organizational levels. In this article, two corporate use cases are used as examples and addressed. In both situations, the massive quantities of information are accelerated. The first is concerned with the knowledge freed from different equipment in the IoT ecosystem. It generates a high quantity of data in a short period. Another example is human-created knowledge using an industrial business system. Section 9.2 discusses the connection between big data and IoT. Section 9.3 discusses big data infrastructure, framework, and technologies. Section 9.4 covers the rationale for and significance of big data. Section 9.5 discusses industrial use cases, operational challenges, methodology, and the importance of data analysis. Section 9.6 discusses several limitations. Section 9.7 brought this chapter to a conclusion.

9.2 The Internet of Things and Big Data Correlation

The Internet of Things is poised to usher with the next industrialization. According to Gartner, revenue produced by IoT devices and related applications would exceed \$3 trillion by 2021. The digitalization using IoT will generate a massive percentage of money and information, and its impact will indeed be felt throughout the world of big data, intriguing enterprises to upgrade existing methods and technology, as well as develop the appropriate advanced technologies to facilitate this increased data volume as well as capitalize on the knowledge and insight from newly conquered data. The massive volume of data generated by IoT would indeed be meaningless without the analytic capability of big data. The Internet of Things and big data are inextricably connected by engineering and commerce. No law says IoT and data science must be linked at the groin; nonetheless, it logically follows them as compatible companions since it is useless to run complicated equipment or devices lacking predictive modeling. This necessitates the use of large amounts of data related to predictive data science for analytics. The "enormous growth of datasets" caused by IoT necessitates the use of big data. Without the finest data collection, companies cannot evaluate the data freed by sensors. Machine and device data are frequently in a basic and simplistic manner; to be used for quantitative choices, the data must be further organized, processed, and supplemented.

9.3 Design, Structure, and Techniques for Big Data Technology

Data analysis, like big data, is defined by three main traits: quantity, speed, and diversity. There seems to be little question that knowledge will continue to be created and acquired, resulting in an amazing volume of data. Furthermore, this data is now being acquired in live time and at a high rate. This is really a sign of speed. Third, various sorts of information are collected in standardized formats and maintained in workbooks or database systems. Addressing the data captured in terms of volume, velocities, and variation, the analytic approaches have evolved to accommodate these characteristics in order to further expand to the sophisticated and nuanced analytics required. Another fourth quality has been proposed by several scholars and researchers: truthfulness. As a result, data integrity is achieved. As a result, the acquired business intelligence tools are extremely trustworthy and error-free. Data analytics is not the same as standard business intelligence technologies. The effectiveness of business intelligence is determined by its infrastructure, instruments, techniques, and methodologies. The Atmospheric & Oceanographic Administering body of the United States uses big data analytics for assist with meteorological conditions & atmospheric surroundings, pattern discovery, and conventional operations.

Data analysis is used by the US Space Agency NASA for its aeronautical and other kinds of research.in the banking industry for investments, loans, customer experiences, and so on. Data analysis is also being used for research by financial, medical, and entertainment firms. To capture and utilize the possibilities of business intelligence, challenges relating to design and infrastructure, resources, techniques, and connections must be resolved. The fundamental infrastructure of big data and analytics is visualized in Figure 9.1. The very first row displays several sorts of large data providers. The information can come from both various sources, and it can be in a variety of forms and locations in a variety of classic and non-traditional processes. All of this information must be gathered for analytics purposes. The original data that had been obtained needed to be converted. Various kinds of sources, in general, release huge data. The upper vertical according to the above structure represents the various services which are used to query, acquire, and analysis the information.

A database engine collects information from diverse sources and makes it accessible for the further investigation. The next one described several big data analysis and infrastructures. The number of mainstream available data is fully accessible. The last one is a representation of the methodologies employed in big data and analytics. Inquiries, summaries, online data analysis preparation (OLAP), and text mining are all part of it. The key result of the overall data science approach is visualization. To gather, process, analyze, and display big data, several approaches and systems have been utilized and developed. Such methods and approaches come from a variety of disciplines.

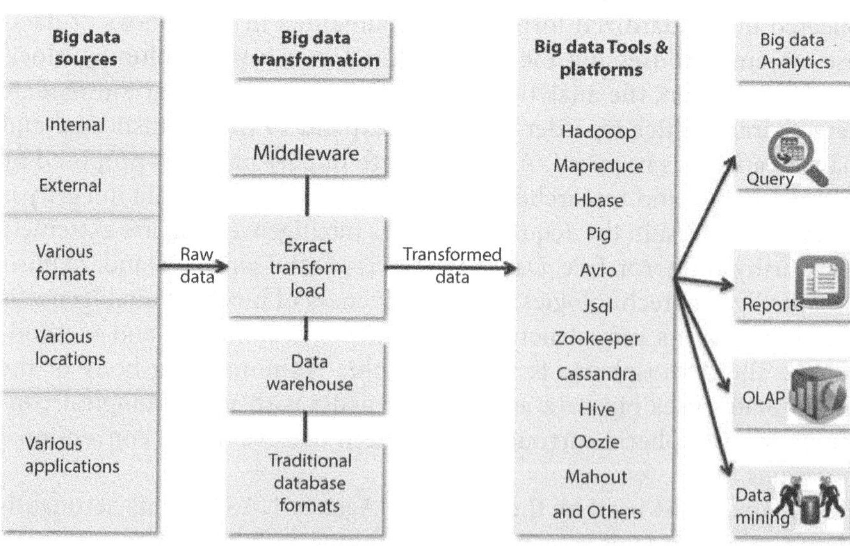

Figure 9.1 Architecture for large-scale data computing in standard.

9.4 Aspiration for Meaningful Analyses and Big Data Visualization Tools

Data science does not simply imply a gradual shift from conventional data processing; it also includes the appropriate real-time business intelligence and visualization tools, as well as the capability to automatically incorporate with conventional networks that are required for business assistance programs, business process management, marketing automation, and decision support systems. Information from disparate data analytic bridge the gap among conventional networks and big data to produce critical results. Consumers' abnormalities, customer support, and online marketing are all examples of smart intelligence. In the end, this one will strengthen the user experience with their merchandise. Private citizens with experience have done well enough in recent years to get into the corporate world. In today's modern circumstances, the legitimate business professionals identify greater insight to reduce company value from huge amounts of data. Business intelligence will assist them in choosing a superior esteem suited for generating the finest company analysis results.

The Internet of Things (IoT) is becoming increasingly important in facilitating access to various devices and equipment in the commercial setting. This change propels us toward digitalization. With the help of IoT, the conventional manufacturing model is transformed into a much more innovative and reliable manufacturing environment. The primary new strategy against an intelligent production facility is to facilitate communication among today's modern external companions with the ultimate objective of relating including an IoT-based production architecture. The IoT-based approach asserts the pyramidal and controlled industrial automation hierarchy by allowing the aforementioned participants to monitor respective services to different layer flattened production environments [1]. It means that the architecture can continue to operate in a shared natural setting rather than in an entangled and significantly linked manner. The interconnected physical environment offers a framework for the creation of novel applications. Organizations are attempting to get even more insights from data by utilizing business intelligence, cloud infrastructures, and a variety of other techniques. Significant challenges associated with the technologically paradigm include rationality, network connection, and architectural environment compatibility.

The lack of a standard current approach for production planning leads to custom-made software or the use of a handcrafted procedure. Furthermore, a joining combined assumption of highly nonlinear components and telecommunication systems is crucial [1]. The notion of ambience cognition is explored in [2]. The article depicts smart classroom rooms, intelligent college campuses, and associated structures. A TensorFlow K-NN classification technique is described in [3]. The Genomic dataset has 90,000,000 pairings. This information will be utilized in the minimization. The Genetic dataset's disequilibrium data was decreased to correct findings without compromising performance [4], addresses the use of Twitter tweets for meta descriptions sentiment analysis. This approach was developed to provide a better knowledge of consumer inclinations. It will aid in advertising strategies and strategic directions. Facebook online data creates a large amount of information. Another sophisticated Fb-Mapping technology [5] has been developed to oversee Facebook data. Emotional responses are unnecessary and hazardous to excellent logic and common sense [6, 7]. A pattern recognition tool [7] is necessary for investigating background and hypotheses emotional states [7]. The document [8] discusses the examination of sociological interpretations based on advances in science and technology with surveillance technology and social scientific studies.

The work proposed by the investigators in respective [9–11] address the use of IoT and machine learning in medical institutions and data analysis. The Economist Intelligence Unit presented a paper a paper [12] considers the implications of exporting production in the region as a whole. The data analyst predicted that they will enter the industrialization, which would focus on industrial digitalization, often known as smart production. The Internet of Things (IoT) is a critical element of industrial automation. Regardless of the fact since M2M communication, digitalization, Scada, PC-based microcontroller, and biosensor usage are all currently in use in various companies, they are mostly disconnected from IT and functional structures. As a result, timely decision making and actions are lacking in many undertakings. Following a role is critical for any organization to push toward the information examination.

9.4.1 From Information to Guidance

Information is only helpful when this is decoded into significant meaningful insights. The great majority of businesses rely on data to make sound decisions. The three critical important factors necessary for persuasive making decision in the commercial environment are the right kind of people, the right moment, and the appropriate facts. Figure 9.2 shows essential judgment call factors necessary in an industrial context. The innermost triangle in the image represents different organizational decisions need to be

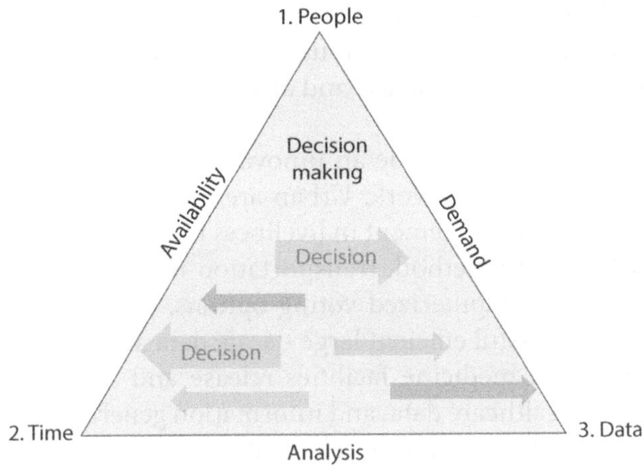

Figure 9.2 Important decision-making considerations.

made. Choices are made more quickly if the appropriate data is provided to the right audience at the right moment. Individuals, knowledge, and opportunity are the three fundamental necessary components.

Accessibility to be recognized to the individuals at the appropriate moment, desire may be calculated from available information and acknowledged to the people in addition, the acquired data had to be evaluated in real - time basis. Analytics-derived insights drive forceful strategic planning. A most successful strategic planning incorporates a smorgasbord of sources of data and provides a comprehensive perspective of the business. Irrelevant data can occasionally become a crucial component in large data. Organizations must understand the critical data relationships that exist among diverse data sources categories.

9.4.2 The Transition from Information Management to Valuation Offerings

As from standpoint of creativity, today's volume of information is a gigantic quantity, continuous knowledge availability, and semi - structured and unstructured content. A reliable data analytics platform should be capable of transforming a large amount of data into meaningful and informative encounters. This will lead to better business decisions.

To fully explore the benefits of business intelligence, the system should be developed with legitimate and analytic applications to facilitate informed decision for continual results from computers. Meaningful data analysis gives significant knowledge into processes. It boosts operational effectiveness. This is very useful for performance monitoring and management software. Big data has been used in a variety of endeavors, and it derives value from a huge database and answers in real time.

1. Smart buildings provide an innovative perspective on how metropolitan areas work. Urban areas are meant to satisfy the pressing management in liveliness requests, preventative social security method, transportation infrastructure, electronic and computerized voting options, etc, which necessitates successful efficient large - scale data administration.
2. Science and medicine facilities release and analyze a vast range of healthcare data, and information generated by diagnostic instruments has accelerated the use of data science. Extensive dataset consists interested in the development of Genomic DNA, diagnostic imaging, molecular characterization, clinical records, and inquiry, among other things. Extracting useful information insight from such a huge data set would assist clinicians in making prompt decisions.
3. Massive developments are taking place in the realm of communications devices. Mobile phone is rising by the day. Huge amount of data are used to derive insights to achieve the greatest amount of network quality by evaluating traffic management, hardware requirements, predicting broken equipment, and so forth.
4. Manufacturing businesses commonly integrate different types of sensors in manufacturing equipment to monitor the effectiveness of the equipment, which aids in the prevention of maintenance issues. The eventual aim of digitalization is to adopt better at each and every stage of the production process. The sensor used is affected by the nature of the activity and the merchandise. As a general rule, delivering the correct information to the correct individual time is a critical component of industrial automation.

9.5 Big Data Applications in the Commercial Surroundings

The first step in realizing the concept of device-to-device communication or intelligent systems is to understand the current production system. The

IoT-based solutions are thought to be capable of transforming the traditional manufacturing configuration into industrial automation. The informational system is an essential transformational component in directing industrial businesses into the next transition. This section represents two usage examples for data science with in manufacturing enterprise. The machinery unified data analytics paradigm is depicted inside one utilization case, while the humanly directed organizational business plan is depicted in the other.

9.5.1 IoT and Data Science Applications in the Production Industry

IoT is an element in the development of digitalization and product improvement. The primary requirement for Industrial revolution 4.0 is the inclusion of IoT-based smart industrial flavors. The information network in the production setup significantly reduces human involvement and allows for automatic control. IoT will assist policymakers in inferring decisions and will maximize efficiency and transparency of manufacturing line statistics. It provides immediate input from the industrial plant's activities. This provides you the opportunity to act quickly if the plan deviates from actuality. This section outlines how the Internet of Things is implemented in the production line. The overall design of sensing connection with machinery is depicted in Figure 9.3.

Its architectural style consists of five phases. Its main stage communicates well with machineries that are linked to various sensors and devices in order to obtain information. The message signal is routed via the central hub. The network communicates via a remote or connected way. The information was then sent to support additional judgment call.

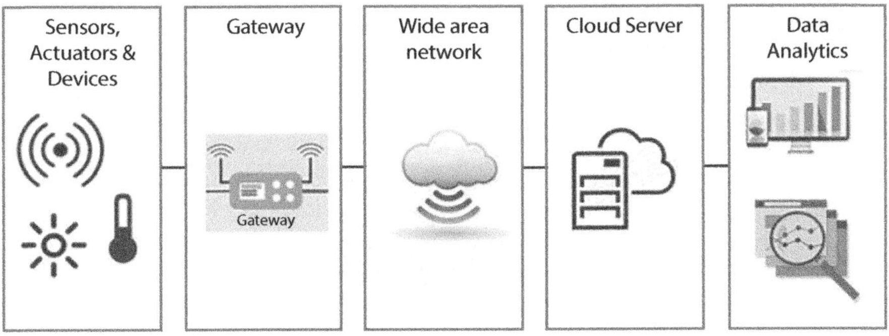

Figure 9.3 To show the overall design of sensing connection with machinery.

The revolutionary advanced analytics platform is indeed the end result of the IoT infrastructure as a whole. This part describes the many strategies used for collection of data as elements of IoT, as well as the translation of information received into the appropriate data structure including data analysis procedures.

Following the adoption of IoT in companies, there has been a huge increase in the number and complexity of data provided by equipment. Examining huge amounts of data reveals a new technique for creating improvement initiatives. Huge data analytics enable the extraction of knowledge through machine-generated large datasets. It provides an opportunity to make companies more adaptable and to respond to calls that have been previously thought to be out of our grasp. With above Figure 9.4 depicts the basic layout of the IoT interconnected in a production-based industrial enterprise.

The first process is to establish a sensor system with instruments. An effective information analytics platform created and implemented to enable employees at all organizational levels to produce better quality decisions based on available information collected from several systems. The procedures that go along with information processing are incorporated in the given data analytic system. The overall structure is divided into three key stages. All of the important steps are covered in detail here.

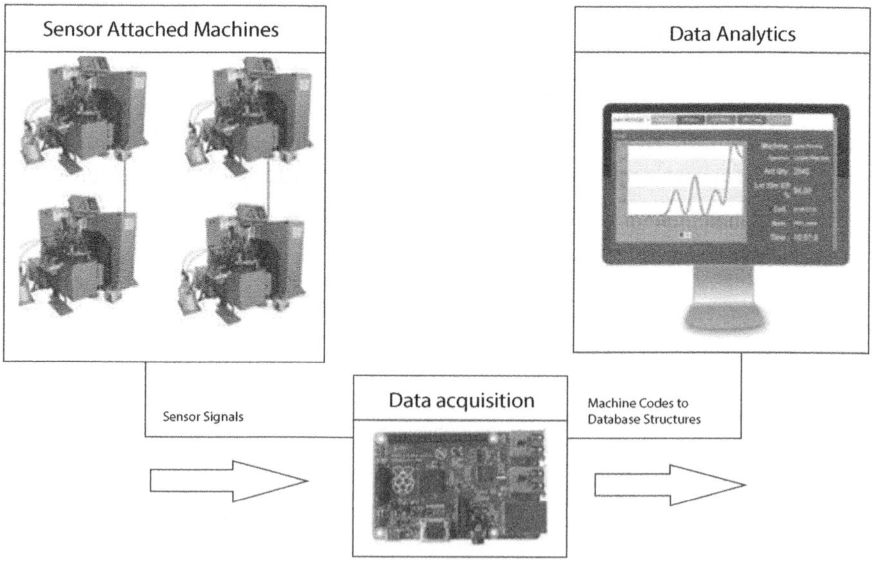

Figure 9.4 To show the basic layout of the IoT interconnected in a production-based industrial enterprise.

9.5.1.1 Devices that are Inter Linked

A scanner is a device that transforms physical parameters into equivalent electrical impulses. Sensors are chosen depending on the attributes and kinds of commodities, operations, and equipment. Many probes are commercially available, such as a thermal imaging sensor, a Reed gauge, a metal gauge sensor, and so on. Preferential sensors may then be linked to machinery depending on the information gathering requirements. The impulses sent by machineries are then routed to the acquisition system. Each instrument connected to the detectors is designated as a distinct cluster. The information gathered from sensors are being sent to the commonly used data collection equipment. Figure 9.5 depicts the information transfer. Each device includes a sensor that converts mechanical characteristic features into electrical impulses.

9.5.1.2 Data Transformation

Data collection is the process of converting physical electrical impulses into binary signals that could be managed by one computing device. It usually converts the signal conditioning impulses supplied by detectors to electronic information for subsequent processing. Figure 9.6 depicts the analogue-to-digital conversion. Managing the information recorded by machinery is a significant problem in the industrial setting. The data

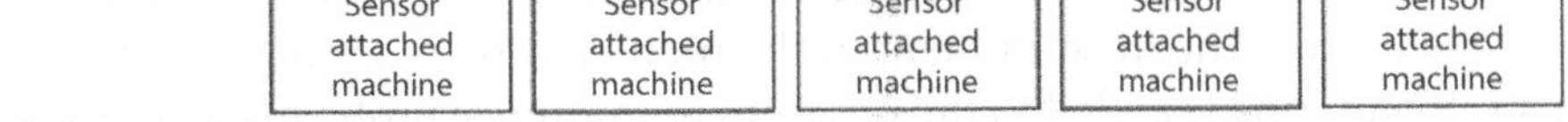

Figure 9.5 Signal transmission from multiple equipment toward a data acquisition system.

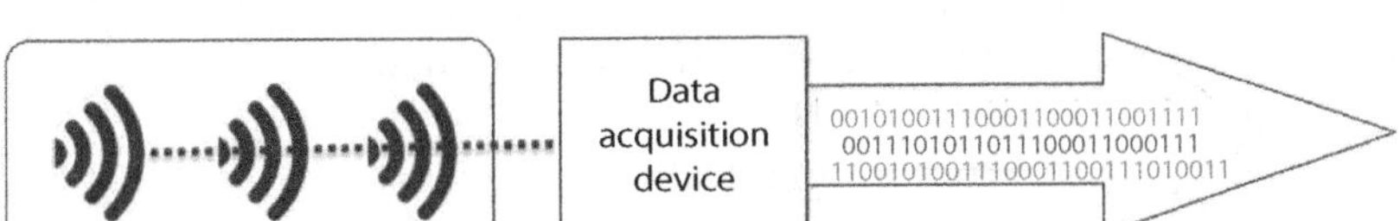

Figure 9.6 To show the analogue-to-digital conversion.

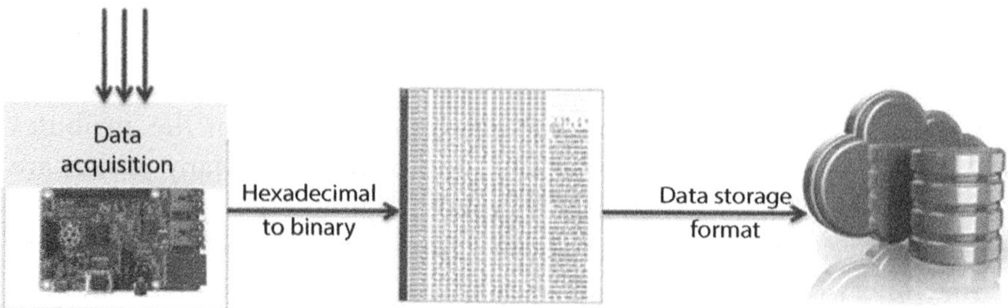

Figure 9.7 To show the overall organization of information acquirer operations.

processing device's source is mechanical impulses, and its return is alphanumeric numbers sent from the acquisition system.

Figure 9.7 depicts the overall organization of information acquirer operations. Data collection is an essential stage in industrial automation. It frees huge amounts of data at incredible speeds. While operating, it transfers data every second. Such information is massive, complicated, and contains a lot of information. Because the dataset is large, an efficient and effective gathering and conversion procedure is necessary. The entire acquiring procedure is conducted in the following steps.

Step I: Information Collection and Storage

This device serves as a bridge among different sensors, as well as a computer architecture. The constant information is transferred from various equipment are the most essential element of the data acquisition system. The interface is in charge of data transport. It gathers information each 20 milliseconds. Application software is used to carry out data collecting activities. These appropriate statistical instructions are product dependent and differ from one instrument to the next. The alphanumeric format was then converted to binary in order to differentiate between active and dormant devices, as well as their respective statuses. Each port corresponds to a single device. The most difficult problem, therefore, is getting the proper data. A buffering is added to the program to minimize problems while data transmission. In certain situations, the information generated by detectors is incomplete and banal. The result of one sensing instrument is not as much as same as the output from others. Legitimate analysis of the data necessitates an accessible and faster processing of data.

Step II: Cleaning and Processing of Data

Typically, the collected data will not be in the suitable form for analysis purposes. The information filtering procedure extracts the important data from the sensor information that is appropriate for analysis. Obtaining the appropriate facts is a technological issue.

Step III: Representing Data
The data processing approach is a difficult process that necessitates a high degree of information unification and consolidation in an autonomous way, ensuring effective and thorough analysis. This procedure necessitates the use of a data model to hold operational data in a system setting.

Step IV: Analytical Input
The IoT-enabled data enables the company to derive meaningful insights with both the assistance of a smart analytical technique. Data analytics assist companies in outfitting existing information and assisting businesses in determining open entryways. This improved research helps the company make, better strategic company actions, better profitable activities, and increased customer retention. The search method is not like in typical database. These data have been collected from the devices in this scenario. Occasionally chaotic data may join the collection of data as a result of environmental disruptions. Detection and eradication of such material is strongly advised in big data [14].

To obtain personal experience from the provided data, query processing options must be used intelligently. It should give actionable concrete answers. At that time, the data gleaned will be stored in a file for subsequent examination. Monitoring is a critical stage for machine-to-machine communication. This is a point of contact between humans and machines. The interface's data should be in a client-acceptable format. Policymakers must understand the graphical types of assessment and extracting meaningful intelligent findings. Figure 9.7 depicts a snapshot of a facility's rotating machines normal operating condition. Every square represents the device's condition.

The white tone indicates that the equipment is operational. The devices indicated by white shade areas are unaffected. The grey hues indicate that the equipment is functioning at a reduced production potential. This increases the performance of both the operator and the device. The dark shade tiles represent the device's idle condition. The idle state could be intentional or unintentional. The image below is from a huge monitor at the production factory. So that everyone in the plant is aware of the immobility and can act immediately. The idle situation is addressed for that in Figure 9.8 for the devices M12, device M24, device M42, and device M44. The devices M21, device M32, device M31, and device M46 are grey, indicating that they are functioning at a low performance. There were no difficulties detected in any of the remaining white-shaded devices. The standard machine condition obtained in Figure 9.8 is displayed on a larger screen at the production site.

Each device's operational condition may be viewed by personnel in the production plant. By this transparency, action may be taken immediately. This causes the manufacturing team to move quickly. A good visualization structure will communicate the best results of the queries in a more understandable manner. Figure 9.9 shows a sample snapshot of the device's current condition.

This graph represents the device's inactivity and operating status. This is an outcall supervisory to make rapid actions. It displays not only the device's current state, as well as the device's operating condition. The starting numbers in Figure 9.9 were inactive in the chart, and the condition altered when the device began. The chart's history clearly demonstrates this. This display is handy for viewing the overall pattern of all devices. Figure 9.10 depicts the operational state of a single console.

The single processor condition displays extra information such as the product title, amount of output, and overall equipment effectiveness. The display pictures used here are examples of the architecture. This device may provide a variety of outcomes. With aid of legitimate dataset, predictive modeling ought to be feasible, so repair actions ought to be able to begin shortly after getting the incorrect signals from the device. By transferring

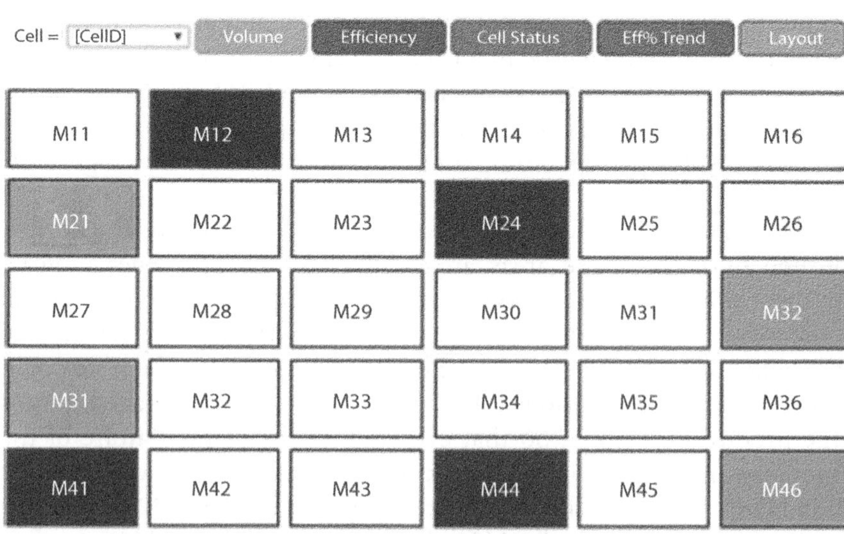

Figure 9.8 To show the standard machine condition.

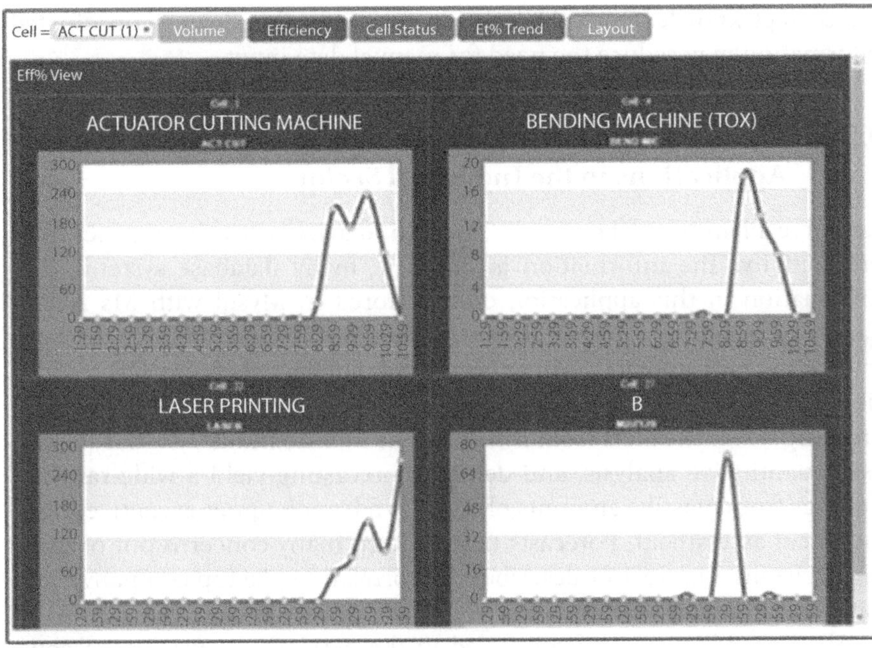

Figure 9.9 To show the overall working efficiency of the production devices.

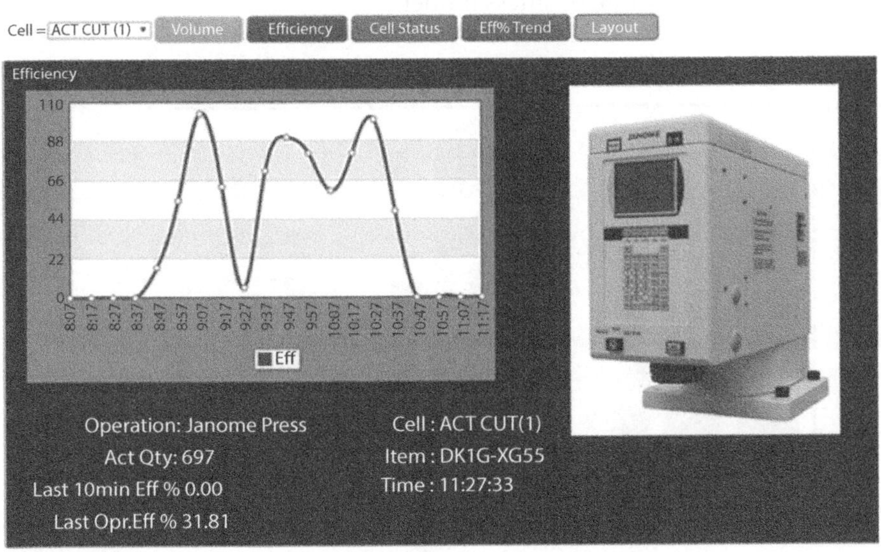

Figure 9.10 To show the operating condition of every individual device.

the correct knowledge to the existing structure, this device coordinated information may reduce the need for manual data input.

9.5.2 Predictive Analysis for Corporate Enterprise Applications in the Industrial Sector

Resource planning data may be used to analyze sales, inventories, and productivity. The information is stored in many database systems. The information in this application case is stored in Mysql with Ms Access databases. All have a diverse data structure. In the event of large datasets, integrating both and delivering meaningful intelligence is a critical responsibility. Many what-if analyses are quite beneficial in comprehending and breaking down the facts from inside. In the finance industry, what-if analysis. quantitative analysis, and demand forecasting yield a wide range of findings from massive amounts of data. To advance, upper executives need judgment assessment. Forecasts in time keep many concerns out of making decisions. Figure 9.11 describes the correlation of a top company's revenues per year ago vs in the year.

All client identities are concealed in the network for privacy reasons. But rather one of measuring the data with some columns, this analytics platform successfully obtains the areas of data. Figure 9.12 depicts product-specific revenues. It generates a graph based on the data collected in the automated corporate business model.

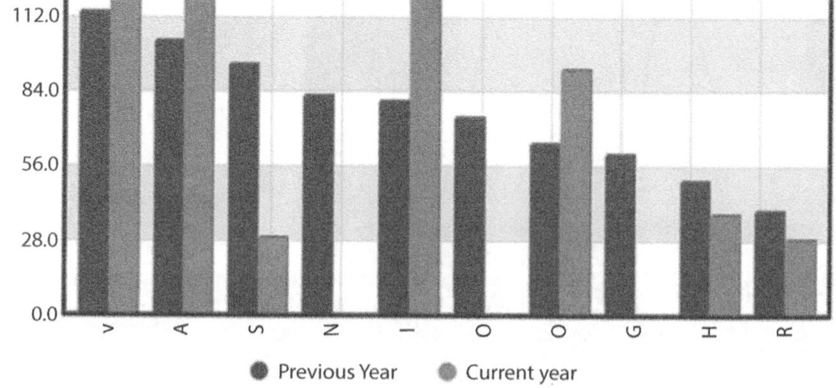

Figure 9.11 To show the correlation of a top company's revenues per year ago vs in the year.

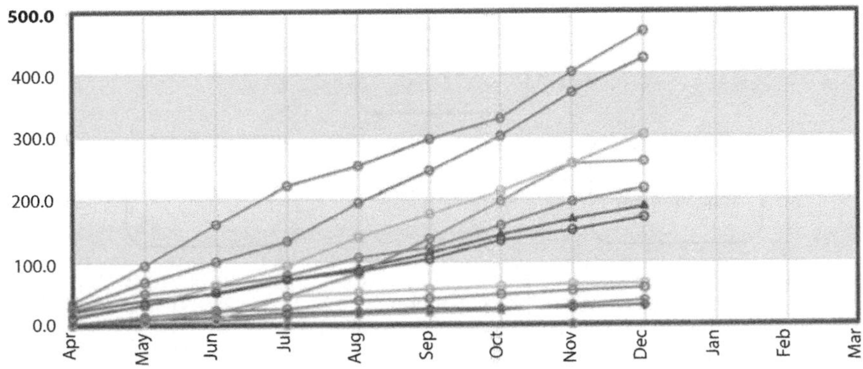

Figure 9.12 To show the product-specific revenues.

This aids in comprehending the demand for various market segments. Organizations can opt for better choices in product categories in which more precision of thought is required. Figure 9.13 below depicts a pay period inventory check. This computational modeling was obtained from of the corporate business program's accessible values. The corporate system comprises a massive dataset including all of the firm's information. Computational modeling is used to route and visualize data that is useful for corporate decision making. Figures 9.11 and 9.12 depict marketing and customer data, whereas Figure 9.13 depicts material-related data. This information is available on the cloud.

As a result, businesses may observe and make decisions from any location. For both situations, intelligence assists upper managers in formulating timely and major decisions. Figure 9.14 depicts the many systems that comprise a conventional manufacturing business.

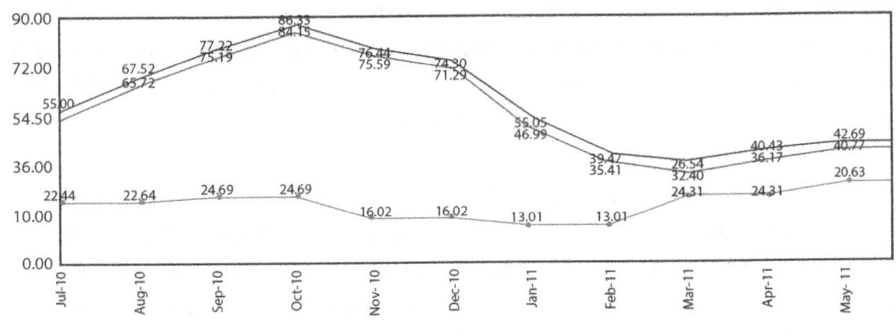

Figure 9.13 To show material-related data available on the cloud.

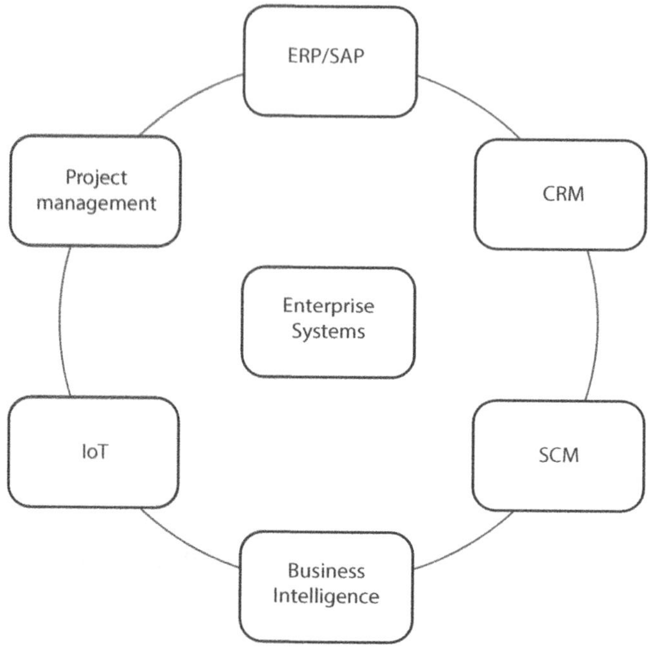

Figure 9.14 To show systems that comprise a conventional manufacturing business.

A cloud-based enterprise resource or system services and commodities framework assists businesses in managing critical aspects of their operations. It combines all of the company's current business operations into a single structure. The integration of organizational resource planning and supply chain operations improves the whole distribution network in the production industry. The arrival of IoT and advancements in computer technology give a much more significant opportunity to develop stronger client interactions. Each organization's overall business goal is to increase revenue; the Customer Connection Management software opens the possibility of a better experience while decreasing communication overhead.

The Costumer Connection Management software paves the change for a successful connection while reducing communication costs. Resource Chain Management is the accomplishment of the connection with the objective of exchanging more loyal consumer value of a product in the production. IoT application scenarios are a clever way of gathering input. The IoT connection does not need human involvement. It collects data from of the device periodically. Data analysis is a way of extracting, modifying, analyzing, and organizing huge amounts of data using computational equations to generate knowledge and information that can be used to make important selections. Despite the fact that business intelligence provides a lot of useful information from a large number of data, it also has certain problems. These difficulties are addressed in the following section.

9.6 Big Data Insights' Constraints

Managing massive amounts of data is the most difficult problem in big data technology. Converting unorganized data into an organized information is a major concern, and afterwards cleansing it first before applying the data analytics is another major issue. Data information available in the traditional model provides data product related, related to a client or a provider, the durability of the material, and so on. Many businesses are taking creative steps to meet the Smart manufacturing requirement. This requirement necessitates the use of IoT. As demonstrated in case study 1, the architecture should be capable of accurately anticipating and assisting individuals in making better decisions in real time. Major companies have started to alter their operations in order to address the difficulties posed by big data.

9.6.1 Technological Developments

The current technique allows for appropriate information storage and retrieval. However, it necessitates a specific attention in the field of IoT and the processing of machine-generated data. Aside from mining techniques, these following steps should be taken:

(a) Formulate appropriate suitable technique and design
(b) Improved first most current application's flexibility and reliability'
(c) Creating commercial value through large datasets.

Merely going towards business intelligence will not assist us till we understand and create economic potential from the long - standing research. Adopting innovative data science tactics, computation offloading, and unique tools will aid in the extraction of relevant insights in businesses. Businesses should indeed be prepared to accept these changes.

9.6.2 Representation of Data

The goal of large datasets is not merely to create a huge collection. It all comes down to producing advanced computers and intelligence. It is important to select a somewhat more suitable business intelligence technology. It ought to, since they visualize the combined information quality produced by the system and comprehension by the device. Its major benefits include

(a) Referred to collectively universal values via given assistance;
(b) To use an autonomous technique to speed computer produced analysis of data.
(c) Dilemma options and the relevance of data gathering.

9.6.3 Data That Is Fragmented and Imprecise

The management of unstructured and structured data is a significant problem in big data. During the troubleshooting step, the production machine ought to be able to interpret how to process the information. In the context of human data usage, variability is easily accepted. Filtering of flawed information is a difficult task in big data technology. Also, after information purification, there is still some tarnished and dirty information in the data collection. Coping with this during the data collection stage is by far the most severe challenge.

9.6.4 Extensibility

For a long period of time, managing large databases with constantly expanding data has been a difficult challenge. Regardless, current developments in network connection, sensing devices, and medical systems are producing massive amounts of data. Initially, this problem was alleviated by the introduction of high-end CPUs, storage systems, and simultaneous data analysis. The next new paradigm shift is upon using the cloud technology, which is based on resource sharing. It is not enough to provide technological platform for data handling. It necessitates a new level of data administration in terms of data preparation, search handling algorithms, database architecture, and fault management mechanisms.

9.6.5 Implementation in Real Time Scenarios

Performance is a critical component in real information execution. The output may be required rapidly in a spectrum of uses. In our first case study, the machine is linked to data gathering equipment for predictive analytics. On the machine, continuous selections such as device shutdown alarms and efficiency are established. So immediate action is necessary upon it. In the event of shopping online, banking transactions, detectors, and so on, rapid execution is necessary. Analyzing the entire data collection to respond the questions on the real time scenario is not feasible.

This problem would be solved by using the appropriate clustering algorithm. Nowadays, a most difficult problem for most businesses is turning mounds of data to findings and then transforming those findings into meaningful commercial benefit. KPMG Worldwide [13] surveyed many leaders in the sectors on real-time analytics. According to the results of the poll, the following are the most significant challenges in business intelligence.

(a) Choosing a corrective solution for precise data analysis.
(b) Identifying appropriate risk factors and measurements.
(c) Movement in real - time basis.
(d) Data analysis is critical.
(e) Offering predictive analytics in all areas of the company.

Whenever technology progresses, it becomes extremely difficult to get meaningful information. However, a new, technological superiority always will arise to forecast market development prospects. Despite the numerous obstacles of big data, each company needs predicted analysis to detect unexpected correlations in massive amounts of data.

9.7 Conclusion

This chapter has discussed the critical functions of data analysis in the industrial sector, namely in the IoT context and as a significant actor in the maneuverable business climate. Most companies' success necessitates the acquisition of new skills and also different perspectives about how to manage big data, which has the potential to accelerate business operations. The modern advanced analytics that have emerged with real liberal company models are an important component of this creative method. The inventive capacities of the rising big data phenomena were explored and dealt in this chapter, as were numerous concerns surrounding its methodology for modifications. The major conclusions are supported by portraying real-life instances.

Various difficulties relating to big data's expand the knowledge and modeling tactics adopted by a number of significant commercial companies. In truth, it is clear that big data is now included into the workflows of several organizations, not because of the buzz it generates transit for its innovative potential to completely transform the business landscape. Although novel big data approaches are always emerging, we been capable of covering a few major ones that are paving the way for the development of goods and services for so many businesses. We are living in the age of big data.

A data-driven business is very effective at forecasting consumer behavior, financial situation, and Supply Chain Management systems. Improved analysis allows businesses to gain deeper information that would increase revenue by delivering the correct goods, which would need more in-depth knowledge. Greater insights would be required for business decisions. The technological problems mentioned in this study must be overcome in order to fully exploit the stored information. Because although data analytics is a strong decision-making resource, information dependability is critical.

In the data and analytics paradigm, there are several possible research avenues. Many governments and industrial companies across the world are shifting their focus to industrial automation in order to attain Industry 4.0. The primary guiding principle for this vision is the concept of technologically operation, in which the manufacturer is heavily connected to be software focused, data driven, and digitized. Total system efficiency is a well-known manufacturing statistic used to offer a gauge of any work center's success. Total system efficiency also provides businesses with a framework for contemplating on IoT application—rebuilding effectiveness, utilization, and reliability.

References

1. Vermesan, O. and Friess, P., *Internet of things- from research and innovation to market deployment*, pp. 74–75, 2014, River Publishers ISBN: 978-87-93102-94-1
2. Bureš, V., Application of ambient intelligence in educational institutions: Visions and architectures. *Int. J. Ambient Comput. Intell.*, 7, 1, 94–120, 2016.
3. Kamal, S., Ripon, S.H., Dey, N., Ashour, A.S., Santhi, V., A MapReduce approach to diminish imbalance parameters for big deoxyribonucleic acid dataset. *Comput. Methods Programs Biomed.*, 131, C, 191–206, 2016.
4. Baumgarten, M., Mulvenna, M., Rooney, N., Reid, J., Keyword-based sentiment mining using Twitter. *Int. J. Ambient Comput. Intell.*, 5, 2, 56–69, 2013.
5. Kamal, S., Dey, N., Ashour, A.S., Ripon, S., Balas, V.E., Kaysar, M.S., FbMapping: An automated system for monitoring Facebook data. *Neural Netw. World*, 27, 1, 27, 2016.
6. Brun, G., Doguoglu, U., Kuenzle, D., Epistemology and emotions. *Int. J. Synth. Emo.*, 4, 1, 92–94, 2013.

7. Alvandi, E.O., Emotions and information processing: A theoretical approach. *Int. J. Synth. Emot.*, 2, 1, 1–14, 2011.
8. Odella, F., Technology studies and the sociological debate on monitoring of social interactions. *Int. J. Ambient Comput. Intell.*, 7, 1, 1–26, 2016.
9. Bhatt, C., Dey, N., Ashour, A.S., Internet of Things and Big Data Technologies for next generation healthcare, Series Title Studies in Big Data, Springer International Publishing, AG, 2017 DOI: https://doi.org/10.1007/978-3-319-49736-5 eBook ISBN 978-3-319-49736-5 Published: 01 January 2017.
10. Kamal, M.S., Nimmy, S.F., Hossain, M., II, Dey, N., Ashour, A.S., Santhi, V., ExSep: An exon separation process using neural skyline filter, in: *International conference onelectrical, electronics, and optimization techniques (ICEEOT)*, 2016, doi: 10.1109/ICEEOT.2016.7755515.
11. Zappi, P., Lombriser, C., Benini, L., Tröster, G., Collecting datasets from ambient intelligence environments. *Int. J. Ambient Comput. Intell.*, 2, 2, 42–56, 2010.
12. *Building Smarter Manufacturing With The Internet of Things (IoT)*, Lopez Research LLC2269, Chestnut Street 202 San Francisco, CA 94123 T(866) 849–5750W, Jan 2014, www.lopezresearch.com.
13. *Going beyond the data: Achieving actionable insights with data and analytics*, KPMG Capital, https://www.kpmg.com/Global/en/IssuesAndInsights/ArticlesPublications/Documents/going-beyond-data-and-analytics-v4.pdf [Date:11/11/2021].
14. Swetha, K.R. and N. M, A. M. P and M. Y. M, Prediction of pneumonia using big data, deep learning and machine learning techniques. *2021 6th International Conference on Communication and Electronics Systems (ICCES)*, pp. 1697–1700, 2021, doi: 10.1109/ICCES51350.2021.9489188.

10

Generative Adversarial Networks: A Comprehensive Review

Jyoti Arora[1*], Meena Tushir[2], Pooja Kherwa[3] and Sonia Rathee[3]

[1]*Department of Information Technology, Maharaja Surajmal Institute of Technology, GGSIPU, New Delhi, India*
[2]*Department of Electronics and Electrical Engineering, Maharaja Surajmal Institute of Technology, GGSIPU, New Delhi, India*
[3]*Department of Computer Science and Engineering, Maharaja Surajmal Institute of Technology, GGSIPU, New Delhi, India*

Abstract
Generative Adversarial Networks (GANs) have gained immense popularity since their introduction in 2014. It is one of the most popular research area right now in the field of computer science. GANs are arguably one of the newest yet most powerful deep learning techniques with applications in several fields. GANs can be applied to areas ranging from image generation to synthetic drug synthesis. They also find use in video generation, music generation, as well as production of novel works of art. In this chapter, we attempt to present detail study about the GAN and make the topic understandable to the readers of this work. This chapter presents an extensive review of GANs, their anatomy, types, and several applications. We have also discussed the shortcomings of GANs.

Keywords: Generative adversarial networks, learning process, computer vision, deep learning, machine learning

List of Abbreviations

Abbreviation	Full Form
GAN	Generative Adversarial Network

**Corresponding author*: jyotiarora@msit.in

M. Niranjanamurthy, Kavita Sheoran, Geetika Dhand, and Prabhjot Kaur (eds.) Data Wrangling: Concepts, Applications and Tools, (213–234) © 2023 Scrivener Publishing LLC

DBM	Deep Boltzmann Machine
DBN	Deep Belief Network
VAE	Variational Autoencoder
DCGAN	Deep Convolutional GAN
cGAN	conditional GAN
WGAN	Wasserstein GAN
LSGAN	Least Square GAN
INFOGAN	Information Maximizing Generative Adversarial Network
ReLU	Rectified Linear Unit
GPU	Graphics Processing Unit

10.1 Introductıon

Generative Adversarial Networks (GANs) are an emerging topic of interest among today's researchers. A large proportion of research is being done on GANs as can be seen from the number of research articles on GANs on Google Scholar. The term "Generative Adversarial Networks" yielded more than 3200 search results for the year 2021 alone (upto 20 March 2021). GANs have also been termed as the most interesting innovation in the field of Machine Learning since past 10 years by Yan LeCun who has a major contribution in the area of Deep Learning networks. The major applications of GANs lie in computer vision [1–5]. GANs are extensively used in the generation of images from text [6, 7], translation of image to image [8, 9], image completion [10, 11].

Ian Goodfellow *et al.* in their research paper "Generative Adversarial Nets" [12] introduced the cocept of GANs. In simplest words GANs are machine learning systems made up of the discriminator, the generator and two neural networks, that generate realistic looking images, video, etc. The generator generates new content which is then evaluated by the discriminator network. In a typical GAN the objective of the generator network is to successively "fool" the discriminator by producing new content that cannot be term as "synthesized" by the discriminator. Such a network can be thought of as analogous to a two player game (zero-sum game i.e the total gain of two players is zero [13]) where the players contest to win. GANs are an adversarial game setting where the generator is pitted against the discriminator [14]. In case of GANs the optimisation process is a minimax game and the goal is to reach Nash equilibrium [15].

Nowedays, GANs are one of the most commonly used deep learning networks. They fall into the class of deep generative networks which also include Deep Belief Network (DBN), Deep Boltzmann Machine (DBM) and Variational Autoencoder (VAE) [16]. Recently GANs and VAE have become popular techniques for unsupervised learning. Though originally intended for unsupervised learning [17–19]. GANs offer several advantages over other deep generative networks like VAE such as the ability of GANs to handle missing data and to model high dimensional data. GANs also have the ability to deliver multimodal outputs (Multiple feasible answers) [20]. In general, GANs are known to generate fine grained and realistic data whereas images generated by VAE tend to be blurred. Even though GANs offer several advantages they have some shortcomings as well. Two of the major limitations of GANs are that they are difficult to train and not easy to evaluate. It is difficult for the generator and the discriminator to attain the Nash equilibrium at the time of training [21] and difficult for the generator to learn the distribution of full datasets completely (leads to mode collapse). The term, mode collapse defines a condition wherein the limited amounts of samples are generated by the generator regardless of the input.

In this paper, we have extensively reviewed Generative Adversarial Networks and have discussed about the anatomy of GANs, types of GANs, areas of applications as well as the shortcomings of GANs.

10.2 Background

To understand GANs it is important to have some background of supervised and unsupervised learning. It is also required to understand generative modelling and how it differs from discriminative modelling. In this section, we attempt to discuss these.

10.2.1 Supervised vs Unsupervised Learning

A supervised learning process is carried by training of a model using a training dataset which consists of several samples with input as well as output labels corresponding to those input values. The model is trained using these samples and the end goal is that the model is able to predict the output label for an unseen input [22]. The objective is basically to train a model in order to generate a mapping capability between inputs, x and outputs, y given multiple labeled input-output pairs [23].

Another type of learning is where a data is given only with input variables (x). This problem does not have labeling of data [23]. The model is built by extracting patterns in the input data. Since the model in question here does not predict anything, no corrections take place here as in case of supervised learning. Generative modelling is a notable unsupervised learning problem. GANs are an example of unsupervised learning algorithms [12].

10.2.2 Generative Modeling vs Discriminative Modeling

Deep Learning models can be characterised into two types—generative models and discriminative models. Discriminative modelling is the same as classification in which we focus on evolving a model to forecast a class label, given a set of input-output pairs (supervised learning). The motive for this particular terminology is to design a model that must discriminate the inputs across classes and make a decision of which class the given input belongs to. Alternatively generative models are unsupervised models that summarise the distributions of inputs and generate new examples [24]. Really good generative models are able to make samples that are not only accurate but also not able to differentiate from the real examples supplied to the model.

In past few years, generative models have seen a significant rise in popularity, specially Generative Adversarial Networks (GANs) which have rendered very realistic results (Figure 10.1). The major difference between generative and discriminative models is that the aim in case of discriminative models is to learn the conditional probability distribution (P(y|x)) whereas, a generative model aims to learn the joint probability distribution (P(x,y)) [25]. In contrary to discriminative models, generative models can use this joint probability distribution to generate likely (x,y) samples. One might assume that there is no need of generating new data samples, owing to the abundance of data already available. However, in reality generative

Figure 10.1 Increasingly realistic faces generated by GANs [27].

modelling has several important uses. Generative models can be used for text to image translation [6, 7] as well as for applications like generating a text sample in a particular handwriting fed to the system. Generative models, specifically GANs can also be used in reinforcement learning to generate artificial environments [26].

10.3 Anatomy of a GAN

A GAN is a bipartite model consist of two neural networks; (i) generator and (ii) a discriminator (Figure 10.2). The task of the generator network is to produce a set of synthetic data when fed with a random noise vector. This fixed-length vector is created randomly from a Gaussian distribution and is used to start the generative process. Following the training, the vector contains points that form a compressed representation of the original data distribution. The generator model acts on these points and applies meaning to them.

The task of the discriminator model is to classify the real data from the one generated by the generator. For doing this, it takes two inputs, an instance from the real domain and another one that comes from the set of examples generated by the generator and then labels them as fake or real i.e. 0 or 1 respectively.

These two networks are trained together with the generator generating a collection of samples. Further, these samples are fed to the discriminator along with real examples which classifies them as real or synthetic. With every successful classification, the discriminator is rewarded while

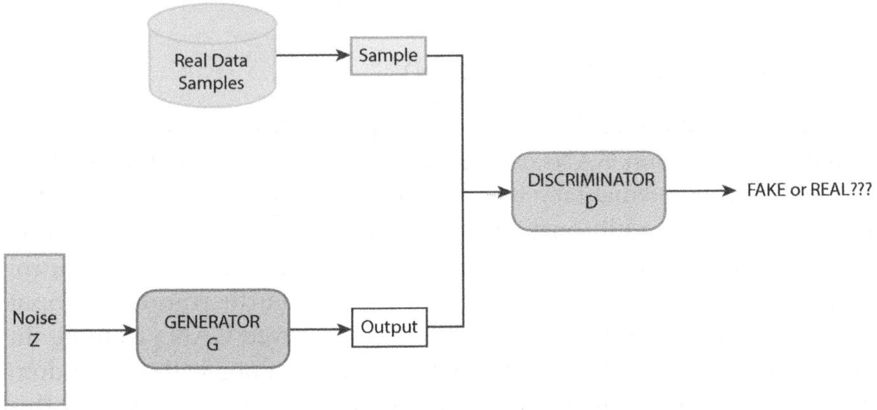

Figure 10.2 Architecture of GAN.

the generator is penalized which it uses to tweak its weights. On the other hand, when the discriminator fails to predict, the generator is rewarded and parameters are not changed while the discriminator is penalized and the parameters of the model are revised. This process continues until the generator becomes skilled enough of synthesizing data which can fool the discriminator or the confidence of correct classification done by the discriminator drops to 50%.

This adversarial training of the two networks makes the generative adversarial network interesting with the discriminator keen on maximizing the loss function while the generator trying to minimize it. The loss function is given below:

$$\min_G \max_D V(D, G) = E_x(\log D(x)) + E_z[\log(1\text{-}D(G(z)))]$$

where, D(x) is the discriminator's confidence, E_x is the expected value over all real data samples, G(z) is the sample generated by the generator when fed with noise z, D(G(z)) is the discriminator's confidence as probability that fake data sample is real and, E_z is the estimated value over all generated fake instances G(z).

10.4 Types of GANs

In this section several types of GANs have been discussed. There are many types of GANs that have been proposed till date. These include Deep Convolutional GANs (DCGAN), conditional GANs (cGAN), InfoGANs, StackGANs, Wasserstein GANs (WGAN), Discover Cross Domain Relations with GANs (DiscoGAN), CycleGANs, Least Square GANs (LSGAN), etc.

10.4.1 Conditional GAN (CGAN)

CGANs or Conditional GANs was developed by Mirza *et al.* [28] with a thought that the plain GANs can be extended to a conditional network by feeding some supplementary information to the generator as well as the discriminator as an additional input layer as shown in Figure 10.3 anything from class labels to data from other modalities. These class labels control the generation of data of a particular class type. Furthermore, the input data with correlated information allows for improved GAN's training. In the generator, the conditional information Y is fed along with the random noise Z merged in a hidden representation while in the discriminator this information is provided along with data instances.

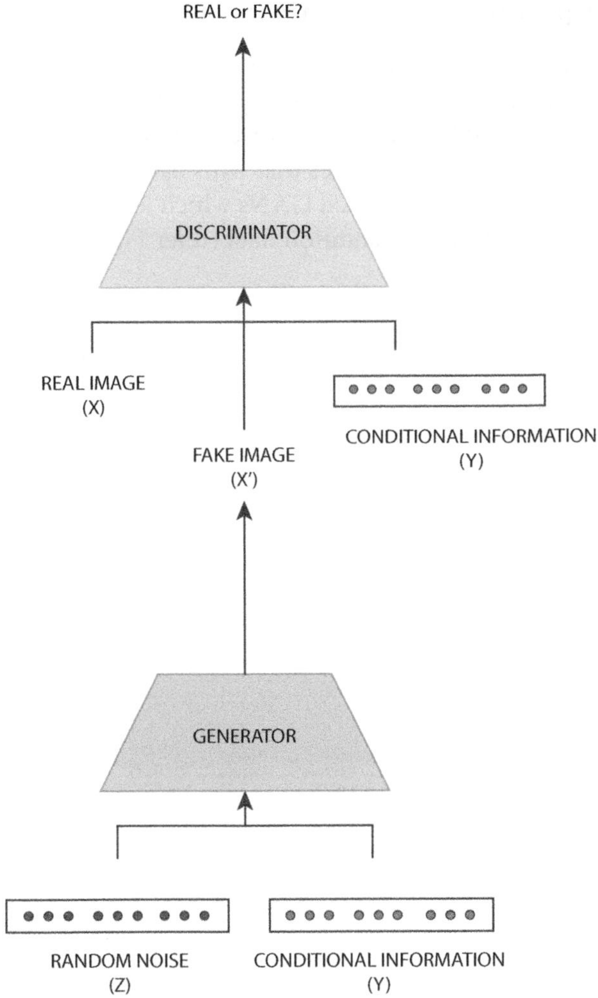

Figure 10.3 Architecture of cGAN.

The authors then trained the network on the MNIST dataset [29] where class labels were conditioned, encoded as one-hot vectors. Building on this, the authors then demonstrated automated image tagging with the predictions using multilabels, the conditional adversarial network to define a set of tag vectors conditioned on image features. A convolutional model inspired from [30] where full Imagenet dataset was pretrained for the image features and for word representation a corpus of text was acquired from the YFCC100M [31] dataset metadata to which proper preprocessing was applied. Finally, the model was then trained on the MIR Flickr dataset [32] to generate automated image tags (refer Figure 10.3).

10.4.2 Deep Convolutional GAN (DCGAN)

These were introduced by Radford *et al.* [33] in late 2015 as a strong contender for practicing unsupervised learning using CNNs in computer vision tasks. The authors of DCGAN mention three major ideas that helped them come up with a class of architectures that wins over the problems faced by prior efforts of building CNN based GANs which lead to training instability when working with high-resolution data (refer Figure 10.4).

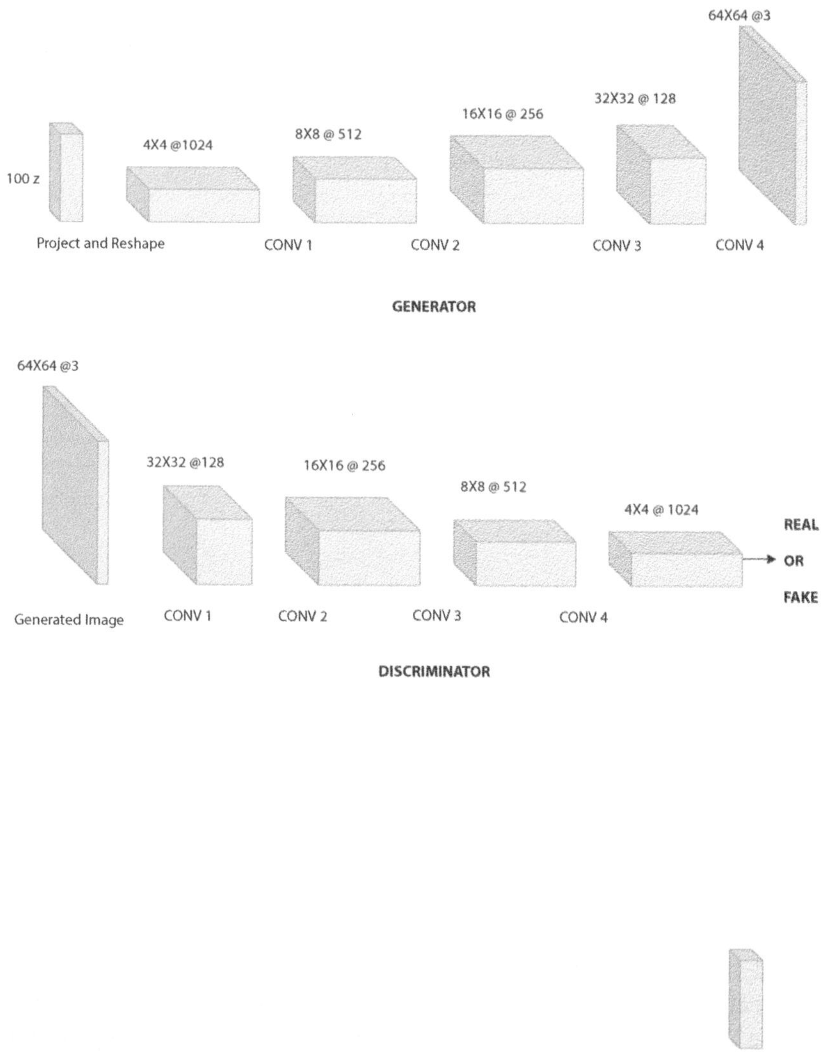

Figure 10.4 DCGAN architecture.

The first idea was to replace any pooling layers with strided convolutional layers in both the discriminator and the generator, taking motivation from the all convolutional network [34]. This allows the network to learn its spatial downsampling. The second was to remove the deeper architectures with fully connected layers and finally, the third idea was to use the concept of Batch Normalization [35] which transforms each input unit to have zero mean and unit variance and stabilizes the learning process by allowing the gradient to flow to deeper models. The technique, however, is not applied to the output layer of the generator and the input layer of the discriminator as its direct application to all the layers leads to training instability and sample oscillations. Additionally, ReLU [36] activation function is used in the generator saving the TanH activation function for the output layer. While the discriminator employs leaky rectified activation [37, 38] which works well with higher resolution images.

DCGAN was trained on three datasets: Large Scale Scene Understanding (LSUN) [39], Imagenet-1k [40] and a then newly assembled faces dataset having 3M images of 10K people. The main idea behind training DCGAN is to use the features realized by the model's discriminator as a feature extractor for the classification model. Radford *et al.* in particular used the concept combined with a L2+SVM classifier which is when tested against the CIFAR-10 dataset leads an 82.8% accuracy.

10.4.3 Wasserstein GAN (WGAN)

They were introduced in 2017 by Martin Arjovsky *et al.* [41] as an alternate to the traditional GAN training methods that had proven to be quite delicate and unstable. WGAN is an impressive extension to GANs that improves stability while the model is being trained as well as helps in analysing the quality of the images generated by associating them with a loss function. The characteristic feature of this model is that it replaces the basic discriminator model with a critic that can be trained to optimality because of the Wasserstein distance [42] which is continuous and differentiable. Wasserstein distance is better than Kullback-Leibler [43] or Jensen-Shannon [44] divergences as it seeks to provide the minimum distance with a smooth and meaningful representation between two data distribution probabilities even when they are located in lower dimensional manifolds without overlaps.

The most compelling feature of WGAN is the drastic reduction of mode dropping phenomenon that is mostly found in GANs. A loss metric is correlated with the generator's convergence. It is backed up by a strong mathematical motivation and theoretical argument. In simpler terms, a reliable gradient of Wasserstein GAN can be obtained by extensively training the critic. However, it might become unstable with the use of momentum-based optimiser (on critic), such as Adam optimizer [45]. Moreover, when the training of the algorithm is done by the generator without constant number of filters and batch normalization, WGAN produces samples while standard GAN fails to learn. WGAN does not show mode collapse when trained with an MLP generator with 4 layers and 512 units with ReLU non-linearities while it can be significantly seen in standard GAN. The benefit of WGAN is that while being less sensitive to model architecture, it can still learn when the critic performs well. WGAN promises better convergence and training stability while generating high quality images.

10.4.4 Stack GAN

Stacked Generative Adversarial Networks (StackGANs) with Conditional Augmentation [46] for synthesizing 256*256 photorealistic images conditioned on text descriptions was introduced by Han Zhang *et al.* [46]. Generating high-quality images from text is of immense importance in applications like computer-aided design or photo-editing. However, a simple addition of unsampling layers in the current state-of-the-art GAN results in training instability. Several techniques such as energy-based GAN [47] or super-resolution methods [48, 49] may provide stability but limited details are added to the images with the low resolution like 64*64 images generated by Reed *et al.* [50].

StackGANs overcame this challenge by decomposing the text-to-image synthesis into a two-stage problem. Stage I GAN sketches follow the primitive shape and basic colour constrained to the given text description and yields a image with the low-resolution. Stage II GAN rectifies the faults in resulting in Stage I by reading the description of the text again and supplements the image by addition of compelling details. A new augmentation technique with proper conditioning encourages the stabilized training of conditional GAN. Images with the more photo realistic details and the diversities are generated using STACK GAN.

10.4.5 Least Square GAN (LSGANs)

Least Square GANs (LSGANs) was given by Xudong Mao, *et al.* in 2016 [51]. LSGANs have been developed with an idea of using the least square loss function which provides a nonsaturating gradient in the discriminator contrary to the sigmoid cross entropy function used by Regular GANS. The loss function based on least squares penalizes the fake samples and pulls them close to the decision boundary. The penalization caused by the least square loss function results to generate the samples by the generator closer to the decision boundary and hence they resemble the real data. This happens even when the samples are correctly seperated by the decision boundary. The convergence of the LSGANs shows a relatively good performance even without batch normalization [6].

Various quantitative and qualitative results have proved the stability of LSGANs along with their power to generate realistic images [52]. Recent studies [53] have shown that Gradient penalty has improved stability of GAN training. LSGANs with Gradient Penalty (LSGANs-GP) have been successfully trained over difficult architectures including 101-layered ResNet using complex datasets such as ImageNet [40].

10.4.6 Information Maximizing GAN (INFOGAN)

Information Maximizing GANs (InfoGAN) was introduced by Xi Chen *et al.* [54] as an extension with the information-theory concept to the regular GANs with an ability to learn disentangled representations in an unsupervised manner.

InfoGAN provides a disentangled representation that represents the salient attributes of a data instance which are helpful for tasks like face and object recognition. Mutual information is a simple and effective modification to traditional GANs. The concept core to InfoGAN is that a single unstructured noise vector is decomposed into two parts, as a source of incompressible noise(z) and latent code(c). In order to discover highly semantic and meaningful representations the common facts between generated samples and latent code is maximised with the use of variational lower bound. Although there have been previous works to learn disentangled representations like bilinear models [55], multiview perception [56], disBM [57] but they all rely on supervised grouping of data. InfoGAN does not require supervision of any kind and it can disentangle both discrete and continuous latent factors unlike hossRBM [58] which can be useful only for discrete latent variables with an exponentially increasing computational cost.

InfoGAN can successfully disentangle writing styles from the shapes of digits on the MNIST dataset. The latent codes(c) are modelled with one categorical code (c1) that switches between digits and models discontinuous variation in data. The continuous codes (c2 and c3) model rotation of digits and control the width respectively. The details like stroke style and thickness are adjusted in such a way that the resulting images are natural looking and a meaningful generalisation can be obtained.

Semantic variations like pose from lighting in 3D images, absence or presence of glasses, hairstyles and emotions can also be successfully disentangled with the help of InfoGAN. Without any supervision, a high level of visual understanding is demonstrated by them. Hence, InfoGAN can learn complex representations on complex datasets with superior image quality as compared to previous unsupervised approaches. Moreover, the use of latent code adds up only negligible computational cost on top of a regular GAN without any training difficulty.

The idea to use mutual information can be further applied to other methods like VAE [59], semisupervised learning with better codes [60] and InfoGAN is used as a tool for high dimensional data discovery.

10.5 Shortcomings of GANs

As captivating training a generative adversarial network may sound, it also has its own share of shortcomings when it comes to practicality, with the most significant ones being as follows:

A frequently encountered problem one faces while training a GAN is the enormous computational cost it requires. While a GAN might run for hours, on a single GPU and on a CPU, on the other hand, it may continue to run beyond even a day! Various researchers have come forward with different strategies to minimize this problem, one such being the idea of a building an architecture with effecient memory utilization. Shuanglong Liu *et al.* centered around architecture based on a parameters deconvolution, an FPGA-friendly method [61-63]. Based on a similar approach, A. Yazdanbakhsh *et al.* devised FlexiGan [64], an end-to-end solution, which produces FPGA based accelerator which is highly optimized from a high-level GAN specification.

The output of the discriminator calculates the loss function therefore the parameters are updated fastly. As a result, the convergence of discriminator is faster and this affects the functioning of the generator due to which parameters are not updated. Furthermore, the generator does not converges and thus generative adversarial networks suffers the problem of partial or total mode collapse, a state where in the generator is generating almost indistinguishable outputs for different latent encodings. To address this Srivastava *et al.* suggested VEEGAN [65] which contains a reconstructor network, which maps the data to noise by reversing the action of the generator.. Elsewhere, Kanglin Liu *et al.* proposed a spectral regularization technique (SR-GAN) [66] which balances the spectral distributions of the weight matrices saving them from getting collapse which consequently prevents mode collapsing in GANs.

Another difficulty experienced while developing a generative adversarial network is the inherent instability caused by training both the generator and the discriminator concurrently. Sometimes the parameters oscillate or destabilize, and never seem to converge. Through their work, Mescheder *et al.* [67] presented how training a GAN for absolutely continuous data and generator distributions show local convergence while performing unregularized training over a realistic case of distributions which are not absolutely continuous is not always convergent. Furthermore, by describing some of the regularization techniques put forward they analyze that GAN training with an instance or zero-centered gradient penalties leads to convergence. Another technique that can fix the instability problems of GANs is Spectral Normalization, a particular kind of normalization applied to the convolutional kernels which can greatly improve the training dynamics as shown by Zhang *et al.* through their model SAGAN [68].

An important point to consider is the influence that a dataset may have on the GAN which is being trained on it. Through their work, Ilya Kamenshchikov and Matthias Krauledat [69] demonstrate that how datasets also play a key role in the successful training of a GAN by taking into notice the influence of datasets like Fashion MNIST [70], CIFAR-10 [71] and ImageNet [40]. Also, building a GAN model requires a large training dataset otherwise its progress in the semantic domain is hampered.

Adding further to the list is the problem of the vanishing gradient that crops up during the training if the discriminator is highly accurate thereby, not providing enough information for the generator to make progress. To solve this problem a new loss function Wasserstein loss was proposed in the model W-GAN [41] by Arjovsky *et al.* where loss is updated by a GAN method and the instances are not actually classified by the discriminator. For each sample, a number is received as output. The value of the number need not necessarily be less than one or greater than 0, thus to decide whether the sample is real or fake, the value of threshold value is not 0.5. The training of the discriminator tries to make the output bigger for real instances as compare to fake instances. Working for a similar cause Salimans *et al.* in 2016 [72] proposed a set of heuristics to solve the problem of vanishing gradient and mode collapse among others by introducing the concept of feature matching. Other efforts worth highlighting include improving the WGAN [42] by Gulrajani *et al.* addressing the problems arising due to weight clipping, Fisher GAN [73] suggested by Mroueh and Sercu introduced a constraint dependent on the data to maintain the capacity of the critic to ensure the stability of training, and Improving Training of WGANs [74] by Wei *et al.*

10.6 Areas of Application

Known for revolutionizing the realm of machine learning ever since their introduction, GANs find their way in a plethora of applications ranging from image synthesis to synthetic drug discovery. This section brings to the fore some of the most important areas of application of GANs with each being discussed in detail as below:

10.6.1 Image

Perhaps, some of the most glorious exploits of GANs have surfaced in the field of image synthesis or manipulation. A major advancement in the field of image synthesis came in late 2015 with the introduction of DCGANs by Radford *et al.* [33] capable of generating random images from scratch. In the year 2017, Liqian Ma *et al.* [75] proposed a GANs based architecture that when supplied with an input image, could generate its variants with each having different postures of the element in the input image. Some other notable applications of GANs in the domain of image synthesis and manipulation include Recycle GAN [76], an approach based on data-driven methodology. It is used for transferring the content of one video or photo to another; ObjGAN [77], a novel GAN architecture developed by a team of scientists at Microsoft understands sketch layouts, captions, and based on the wording details are refined; StyleGAN [78], a model Nvidia developed, is capable of synthesizing high-resolution images of fictional people by learning attributes like facial pose, freckles, and hair.

10.6.2 Video

With a video being described as a series of images in motion, the involvement of various state-of-the-art GAN approaches in the domain of video synthesis is no surprise. With DeepMind's proposal of DVDGAN [79], the generation of realistic-looking videos by a model when fed with a custom-tailored dataset is a matter of just a few lines of code and patience. Another noteworthy contribution of GANs in this sector is DeepRay, a Cambridge Consultants' creation. It helps to generate images which are less distorted and more sharper from pictures that have been damaged or had obscured elements. This can be used to get rid of noise in videos too.

10.6.3 Artwork

GANs have the ability to generate more then images and video footage. They are capable of producing novel works of art provided they are supplied with the right dataset. Art-GAN [80], a conditional GAN based network generates images with abstract information like images with a certain art style after being trained on the Wikiart dataset. GauGAN [81] developed by the company can turn rough doodles into photorealistic masterpieces with breathtaking ease and NVIDIA Research has investigated AI-based arts as a deep learning model.

10.6.4 Music

After giving astonishing results when applied to images or videos, GANs are being involved in the field of music generation too. MidiNet [82], a CNN inspired GAN model developed by DeepMind is one such attempt that aims at producing realistic melody from random noise as input. Conditional-LSTM GAN [83] presented by the researchers based at the National Institute of Informatics in Tokyo which learns the latent relationship between the different lyrics and their corresponding melodies and then applies it to generate lyrics conditioned melodies is another effort worth mentioning.

10.6.5 Medicine

Owing to the ability to synthesize images with an unmatched degree of realism and the adversarial training, GANs are a boon for the medical industry. They are frequently used in image analysis, anomaly detection or even for the discovery of new drugs. More recently, the Imperial College London, University of Augsburg, and the Technical University of Munich The model dubbed Snore-GAN [84] is used to synthesize data to fill in gaps in real data. Meanwhile, Schlegl *et al.* suggested an unsupervised approach to detect anomalies relevant for disease progression and treatment monitoring through their discovery AnoGAN [85]. On the drug synthesis side of the equation, LatentGAN [86] an effort by Prykhodko *et al.* integerates a generative adversarial neural network with an autoencoder for de novo molecular design. It can be used with many other applications [89, 90].

10.6.6 Security

With GANs being applied to various domains, it seems the field of security has a lot to gain from them as well. A recently developed machine learning approach to password cracking PassGAN [87] generates password guesses by training a GAN on a list of leaked passwords. Keeping their potential to synthesize plausible instances of data, GANs are being used to make the existing deep learning networks used in cybersecurity more robust by manufacturing more fake data and training the existing deep learning techniques on them. In a similar vein, Haichao *et al.* have come up with SSGAN [88], a new strategy that generates more suitable and secure covers for steganography with an adversarial learning scheme.

10.7 Conclusıon

This paper provides a comprehensive review of generative adversarial networks. We have discussed the basic anatomy of GANs and the various kinds of GANs that have been widely used nowadays. This papers also discusses the various application areas of GANs. Despite the extensive potential, GANs have several shortcomings which have also been discussed. This review of generative adversarial networks extensively covers the basic fundamentals about GANs and will help the readers to gain a good understanding of this famous deep learning network, which has gained immense populatrity recently.

References

1. Regmi, K. and Borji, A., Cross-view image synthesis using conditional GANs. *2018 IEEE/CVF Conference on Computer Vision and Pattern Recognition*, pp. 3501–3510, 2018.
2. Wang, T., Liu, M., Zhu, J., Tao, A., Kautz, J., Catanzaro, B., High-resolution image synthesis and semantic manipulation with conditional GANs. *2018 IEEE/CVF Conference on Computer Vision and Pattern Recognition*, pp. 8798–8807, 2017.
3. Odena, A., Olah, C., Shlens, J., Conditional image synthesis with auxiliary classifier gans, in: *Proceedings of the 34th International Conference on Machine Learning*, JMLR, vol. 70, pp. 2642–2651, 2017.
4. Vondrick, C., Pirsiavash, H., Torralba, A., Generating videos with scene dynamics, in: *Advances in Neural Information Processing Systems*, pp. 613–621, 2016.
5. Zhu, J.-Y., Krähenbühl, P., Shechtman, E., Efros, A.A., Generative visual manipulation on the natural image manifold, in: *European Conference on Computer Vision*, Springer, pp. 597–613, 2016.

6. Reed, S.E., Akata, Z., Yan, X., Logeswaran, L., Schiele, B., Lee, H., Generative adversarial text to image synthesis. *Proc. 33rd Int. Conf. Mach. Learning, PMLR,* 48, 1060–1069, 2016.
7. Xu, T., Zhang, P., Huang, Q., Zhang, H., Gan, Z., Huang, X., He, X., AttnGAN: Fine-grained text to image generation with attentional generative adversarial networks. *IEEE/CVF Conference on Computer Vision and Pattern Recognition*, pp. 1316–1324, 2017.
8. Lin, J., Xia, Y., Qin, T., Chen, Z., Liu, T., Conditional image-to-image translation. *2018 IEEE/CVF Conference on Computer Vision and Pattern Recognition*, pp. 5524–5532, 2018.
9. Choi, Y., Choi, M., Kim, M., Ha, J., Kim, S., Choo, J., StarGAN: Unified generative adversarial networks for multi-domain image-to-image translation. *2018 IEEE/CVF Conference on Computer Vision and Pattern Recognition*, pp. 8789–8797, 2017.
10. Akimoto, N., Kasai, S., Hayashi, M., Aoki, Y., 360-degree image completion by two-stage conditional GANS. *IEEE International Conference on Image Processing (ICIP)*, Taipei, Taiwan, pp. 4704–4708, 2019.
11. Chen, Z., Nie, S., Wu, T., Healey, C.G., Generative adversarial networks in computer vision: A survey and taxonomy. 2018, arXiv preprint arXiv:1801.07632, 2018.
12. Goodfellow, I., Pouget-Abadie, J., Mirza, M., Xu, B., Warde-Farley, D., Ozair, S., Courville, A., Bengio, Y., *Generative Adversarial Networks*(PDF). *Proceedings of the International Conference on Neural Information Processing Systems (NIPS 2014)*, pp. 2672–2680, 2014.
13. Wang, K., Gou, C., Duan, Y., Lin, Y., Zheng, X., Wang, F., Generative adversarial networks: Introduction and outlook. *IEEE/CAA J. Autom. Sin.*, *4*, 588–598, 2017.
14. Grnarova, P., Levy, K.Y., Lucchi, A., Hofmann, T., Krause, A., *An online learning approach to generative adversarial networks*, 2017, ArXiv, abs/1706.03269.
15. Ratliff, L.J., Burden, S.A., Sastry, S.S., Characterization and computation of local Nash equilibria in continuous games, in: *Proc. 51st Annu. Allerton Conf. Communication, Control, and Computing (Allerton)*, Monticello, IL, USA, pp. 917–924, 2013.
16. Pouyanfar, S., Sadiq, S., Yan, Y., Tian, H., Tao, Y., Reyes, M.E., Shyu, M., Chen, S., Iyengar, S.S., A survey on deep learning: Algorithms, techniques, and applications. *ACM Comput. Surv.*, *51*, 92, 1–92:36, 2018.
17. Kumar, A., Sattigeri, P., Fletcher, P.T., *Improved semi-supervised llearning with GANs using manifold invariances*, NIPS, 2017, ArXiv, abs/1705.08850.
18. Odena, A., *Semi-supervised learning with generative adversarial networks*, 2016, ArXiv, abs/1606.01583.
19. Lecouat, B., Foo, C.S., Zenati, H., Chandrasekhar, V.R., Semi-supervised learning with GANs: Revisiting manifold regularization. 2018. ArXiv, abs/1805.08957.

20. Goodfellow, I., *Nips (2016) tutorial: Generative adversarial networks*, p. 215, NIPS, arXiv preprint arXiv:1701.00160.
21. Farnia, F. and Ozdaglar, A.E., *GANs may have no nash equilibria*, 2020, ArXiv, abs/2002.09124.
22. Akinsola, J.E.T., Supervised machine learning algorithms: Classification and comparison. *Int. J. Comput. Trends Technol. (IJCTT)*, 48, 128 - 138, 2017. 10.14445/22312803/IJCTT-V48P126.
23. Murphy, K.P., *Machine Learning: A Probabilistic Approach*, p. 216, The MIT Press, 2012.
24. Bishop, C.M., *Pattern Recognition and Machine Learning*, p. 216, Springer, 2011.
25. Liu, B. and Webb, G.I., Generative and discriminative learning, in: *Encyclopedia of machine learning*, C. Sammut and G.I. Webb (Eds.), Springer, Boston, MA, 2011.
26. Kasgari, A.T., Saad, W., Mozaffari, M., Poor, H.V., *Experienced deep reinforcement learning with generative adversarial networks (GANs) for model-free ultra reliable low latency communication*, 2019, ArXiv, abs/1911.03264.
27. Brundage, M., Avin, S., Clark, J., Toner, H., Eckersley, P., Garfinkel, B., Dafoe, A., Scharre, P., Zeitzoff, T., Filar, B., Anderson, H.S., Roff, H., Allen, G.C., Steinhardt, J., Flynn, C., Beard, S., Belfield, H., Farquhar, S., Lyle, C., Crootof, R., Evans, O., Page, M., Bryson, J., Yampolskiy, R., Amodei, D., *The malicious use of artificial intelligence: Forecasting, prevention, and mitigation*, 2018, ArXiv, abs/1802.07228.
28. Mirza, M. and Osindero, S., *Conditional generative adversarial nets*, 2014, ArXiv, abs/1411.1784.
29. Chen, F., Chen, N., Mao, H., Hu, H., *Assessing four neural networks on handwritten digit recognition dataset (MNIST)*, 2018, ArXiv, abs/1811.08278.
30. Krizhevsky, A., Sutskever, I., Hinton, G.E., Imagenet classification with deep convolutional neural networks. *NIPS*, 2012.
31. *Yahoo flickr creative common 100m*, p. 219, Dataset, http://webscope.sandbox.yahoo.com/catalog.php?datatype=i&did=67.
32. Huiskes, M.J. and Lew, M.S., The mir flickr retrieval evaluation, in: *MIR '08: Proceedings of the 2008 ACM International Conference on Multimedia Information Retrieval*, New York, NY, USA, ACM, 2008.
33. Radford, A., Metz, L., Chintala, S., *Unsupervised representation learning with deep convolutional generative adversarial Networks*, 2015, CoRR, abs/1511.06434.
34. Springenberg, J.T., Dosovitskiy, A., Brox, T., Riedmiller, M.A., *Striving for simplicity: The all convolutional net*, 2014, CoRR, abs/1412.6806.
35. Ioffe, S. and Szegedy, C., *Batch normalization: Accelerating deep network training by reducing internal covariate shift*, 2015, ArXiv, abs/1502.03167.
36. Nair, V. and Hinton, G.E., Rectified linear units improve restricted Boltzmann machines. *ICML*, 2010.

37. Maas, A.L., *Rectifier nonlinearities improve neural network acoustic models*, 2013.
38. Xu, B., Wang, N., Chen, T., Li, M., Empirical evaluation of rectified activations in convolutional network, 2015. ArXiv, abs/1505.00853.
39. Yu, F., Zhang, Y., Song, S., Seff, A., Xiao, J., *LSUN: Construction of a large-scale image dataset using deep learning with humans in the loop*, 2015, ArXiv, abs/1506.03365.
40. Deng, J., Dong, W., Socher, R., Li, L., Li, K., Li, F., ImageNet: A large-scale hierarchical image database. *2009 IEEE Conference on Computer Vision and Pattern Recognition*, pp. 248–255, 2009.
41. Arjovsky, M., Chintala, S., Bottou, L., *Wasserstein GAN*, 2017, ArXiv, abs/1701.07875.
42. Gulrajani, I., Ahmed, F., Arjovsky, M., Dumoulin, V., Courville, A.C., Improved training of Wasserstein GANs. *NIPS*, 2017.
43. Ponti, M., Kittler, J., Riva, M., Campos, T.E., Zor, C., A decision cognizant Kullback-Leibler divergence. *Pattern Recognit.*, *61*, 470–478, 2017.
44. Nielsen, F., *On a generalization of the Jensen-Shannon divergence and the JS-symmetrization of distances relying on abstract means*, 2019, ArXiv, abs/1912.00610.
45. Kingma, D.P. and Ba, J., Adam: A method for stochastic optimization, 2014. CoRR, abs/1412.6980, https://arxiv.org/pdf/1412.6980.pdf.
46. Zhang, H., Xu, T., Li, H., StackGAN: Text to photo-realistic image synthesis with stacked generative adversarial networks. *2017 IEEE International Conference on Computer Vision (ICCV)*, pp. 5908–5916, 2016.
47. Zhao, J.J., Mathieu, M., LeCun, Y., *Energy-based generative adversarial network*, 2016, ArXiv, abs/1609.03126.
48. Sønderby, C.K., Caballero, J., Theis, L., Shi, W., Huszár, F., *Amortised MAP inference for image super-resolution*, 2016, ArXiv, abs/1610.04490.
49. Ledig, C., Theis, L., Huszár, F., Caballero, J.A., Aitken, A., Tejani, A., Totz, J., Wang, Z., Shi, W., Photo-realistic Single image super-resolution using a generative adversarial network. *2017 IEEE Conference on Computer Vision and Pattern Recognition (CVPR)*, pp. 105–114, 2016.
50. Reed, Z.A., Yan, X., Logeswaran, L., Schiele, B., Lee, H., Generative adversarial text-to-image synthesis, 2016. arXiv:1609.04802.
51. Mao, X., Li, Q., Xie, H., Lau, R.Y., Wang, Z., Smolley, S.P., Least squares generative adversarial networks. *2017 IEEE International Conference on Computer Vision (ICCV)*, pp. 2813–2821, 2016.
52. Mao, X., Li, Q., Xie, H., Lau, R.Y., Wang, Z., Smolley, S.P., On the effectiveness of least squares generative adversarial networks. *IEEE Trans. Pattern Anal. Mach. Intell.*, *41*, 2947–2960, 2019.
53. Kodali, N., Hays, J., Abernethy, J.D., Kira, Z., On convergence and stability of GANs. *Artif. Intell.*, 2018. arXiv. https://arxiv.org/pdf/1705.07215.pdf.

54. Chen, X., Duan, Y., Houthooft, R., Schulman, J., Sutskever, I., Abbeel, P., InfoGAN: Interpretable representation learning by information maximizing generative adversarial nets. *NIPS*, 2016.
55. Tenenbaum, J.B. and Freeman, W.T., Separating style and content with bilinear models. *Neural Comput.*, *12*, 1247–1283, 2000.
56. Zhu, Z., Luo, P., Wang, X., Tang, X., Deep learning multi-view representation for face recognition, 2014. ArXiv, abs/1406.6947.
57. Reed, S.E., Sohn, K., Zhang, Y., Lee, H., Learning to disentangle factors of variation with manifold interaction. *ICML*, 2014.
58. Desjardins, G., Courville, A.C., Bengio, Y., Disentangling factors of variation via generative entangling, 2012. ArXiv, abs/1210.5474.
59. Kingma, D.P. and Welling, M., Auto-Encoding Variational Bayes, 2013. CoRR, arXiv:1312.6114, abs/1312.6114.
60. Springenberg, J.T., Unsupervised and Semi-supervised Learning with Categorical Generative Adversarial Networks, 2015. CoRR, abs/1511.06390.
61. Liu, S., Zeng, C., Fan, H., Ng, H., Meng, J., Que, Z., Niu, X., Luk, W., Memory-efficient architecture for accelerating generative networks on FPGA. *2018 International Conference on Field-Programmable Technology (FPT)*, pp. 30–37, 2018.
62. Sulaiman, N., Obaid, Z., Marhaban, M.H., Hamidon, M.N., Design and implementation of FPGA-based systems -A Review. *Aust. J. Basic Appl. Sci.*, 3, 224, 2009.
63. Shawahna, A., Sait, S.M., El-Maleh, A.H., FPGA-based accelerators of deep learning networks for learning and classification: A review. *IEEE Access*, *7*, 7823–7859, 2019.
64. Yazdanbakhsh, A., Brzozowski, M., Khaleghi, B., Ghodrati, S., Samadi, K., Kim, N.S., Esmaeilzadeh, H., FlexiGAN: An end-to-end solution for FPGA acceleration of generative adversarial networks. *2018 IEEE 26th Annual International Symposium on Field-Programmable Custom Computing Machines (FCCM)*, pp. 65–72, 2018.
65. Srivastava, A., Valkov, L., Russell, C., Gutmann, M.U., Sutton, C.A., VEEGAN: Reducing mode collapse in GANs using implicit variational learning. *NIPS*, 2017.
66. Liu, K., Tang, W., Zhou, F., Qiu, G., Spectral regularization for combating mode collapse in GANs. *2019 IEEE/CVF International Conference on Computer Vision (ICCV)*, pp. 6381–6389, 2019.
67. Mescheder, L.M., Geiger, A., Nowozin, S., Which training methods for GANs do actually Converge? *ICML*, 2018.
68. Zhang, H., Goodfellow, I.J., Metaxas, D.N., Odena, A., Self-attention generative adversarial networks, 2019. ArXiv, abs/1805.08318.
69. Kamenshchikov, I. and Krauledat, M., Effects of dataset properties on the training of GANs, 2018. ArXiv, abs/1811.02850.
70. Xiao, H., Rasul, K., Vollgraf, R., Fashion-MNIST: A novel image dataset for benchmarking machine learning algorithms, 2017. ArXiv, abs/1708.07747.

71. Krizhevsky, A., Learning multiple layers of features from tiny images, 2009.
72. Salimans, T., Goodfellow, I.J., Zaremba, W., Cheung, V., Radford, A., Chen, X., Improved techniques for training GANs. *NIPS*, 2016.
73. Mroueh, Y. and Sercu, T., Fisher GAN. *NIPS*, 2017.
74. Wei, X., Gong, B., Liu, Z., Lu, W., Wang, L., Improving the improved training of Wasserstein GANs: A consistency term and its dual effect, 2018. ArXiv, abs/1803.01541.
75. Ma, L., Jia, X., Sun, Q., Schiele, B., Tuytelaars, T., Gool, L.V., Pose guided person image generation, 2017. ArXiv, abs/1705.09368.
76. Bansal, A., Ma, S., Ramanan, D., Sheikh, Y., Recycle-GAN: Unsupervised video retargeting. *ECCV*, 2018.
77. Li, W., Zhang, P., Zhang, L., Huang, Q., He, X., Lyu, S., Gao, J., Object-driven text-to-image synthesis via adversarial training. *2019 IEEE/CVF Conference on Computer Vision and Pattern Recognition (CVPR)*, pp. 12166–12174, 2019.
78. Karras, T., Laine, S., Aila, T., A style-based generator architecture for generative adversarial networks. *2019 IEEE/CVF Conference on Computer Vision and Pattern Recognition (CVPR)*, pp. 4396–4405, 2018.
79. Clark, A., Donahue, J., Simonyan, K., Efficient video generation on complex datasets, 2019. ArXiv, abs/1907.06571.
80. Tan, W.R., Chan, C.S., Aguirre, H.E., Tanaka, K., ArtGAN: Artwork synthesis with conditional categorical GANs. *2017 IEEE International Conference on Image Processing (ICIP)*, pp. 3760–3764, 2017.
81. Park, T., Liu, M., Wang, T., Zhu, J., Semantic image synthesis with spatially-adaptive normalization. *2019 IEEE/CVF Conference on Computer Vision and Pattern Recognition (CVPR)*, pp. 2332–2341, 2019.
82. Yang, L., Chou, S., Yang, Y., MidiNet: A convolutional generative adversarial network for symbolic-domain music generation, 2017. ArXiv, abs/1703.10847.
83. Yu, Y.B. and Canales, S., Conditional LSTM-GAN for melody generation from Lyrics, 2019. ArXiv, abs/1908.05551.
84. Zhang, Z., Han, J., Qian, K., Janott, C., Guo, Y., Schuller, B.W., Snore-GANs: Improving Automatic snore sound classification with synthesized data. *IEEE J. Biomed. Health Inform.*, *24*, 300–310, 2019.
85. Schlegl, T., Seeböck, P., Waldstein, S.M., Schmidt-Erfurth, U., Langs, G., Unsupervised anomaly detection with generative adversarial networks to guide marker discovery. *IPMI*, 2017.
86. Prykhodko, O., Johansson, S.V., Kotsias, P., Arús-Pous, J., Bjerrum, E.J., Engkvist, O., Chen, H., A *de novo* molecular generation method using latent vector based generative adversarial network. *J. Cheminformatics*, 11, 74, 2019.
87. Hitaj, B., Gasti, P., Ateniese, G., Pérez-Cruz, F., PassGAN: A deep learning approach for password guessing, 2019. ArXiv, abs/1709.00440.
88. Shi, H., Dong, J., Wang, W., Qian, Y., Zhang, X., *SSGAN: Secure steganography based on generative adversarial networks*, PCM, p. 228, 2017.

89. Hooda, S. and Mann, S., Examining the effectiveness of machine learning algorithms as classifiers for predicting disease severity in data warehouse environments. *Rev. Argent. Clín. Psicol.*, 29, 233–251, 2020.
90. Arora, J., Grover, M., Aggarwal, K., Augmented reality model for the virtualisation of the mask. *J. Multi Discip. Eng. Technol.*, 14, 2, 2021, 2021.

11

Analysis of Machine Learning Frameworks Used in Image Processing: A Review

Gurpreet Kaur[1] and Kamaljit Singh Saini[2*]

[1]University Institute of Computing, Chandigarh University, Mohali, India
[2]University Institute of Engineering, Chandigarh University, Mohali, India

Abstract
The evolution of the artificial intelligence (AI) has changed the 21st century. Technologically, the advancements are quicker than the predictions. With certain advancements in AI, the field of machine learning (ML) has become the trendiest in this century. ML deals with the science that creates computers, which can learn and perform activities like human beings when we fed data and information into them. These computers do not require explicit programming. In this paper, a general idea of machine leaning concepts is given. It also describes different types of machine learning methods and enlightens the differences between them. It also enlightens the applications and frameworks used with ML for analyzing data.

Keywords: Machine learning basics, types, applications, analysis, wrangling, ML in image processing, frameworks

11.1 Introduction

ML is a type of AI that creates computers that work without explicit programming and have ability to learn. ML is all around us in this modern world. It works on developing computer programs, which can access datasets and execute automatically with detections and predictions. It enables machines to learn from experience continuously. Feeding more data into computer system enables them to improve the results. When trained machines come across to new datasets, these grow, develop, learn, and

**Corresponding author*: sainikamaljitsingh@gmail.com

M. Niranjanamurthy, Kavita Sheoran, Geetika Dhand, and Prabhjot Kaur (eds.) Data Wrangling: Concepts, Applications and Tools, (235–250) © 2023 Scrivener Publishing LLC

change by themselves [1]. The application of machine learning use concept of pattern recognition to provide reliable results.

ML deals with computer programming that can be changed when exposed to new data. Machine can learn its code. The machine is programmed once, every time it encounters some problem, it can solve the problem by its analyzing the learned code. There is no need to program it again and again. It changes its own code according to the new scenarios it discovers. It self-learns whatever has to be learnt according to provided scenarios, past experiences, from provided values and it comes up with new solutions [2, 3]. Here the question arises that "How a machine can recode its code by own?" As per the study, plenty of research has been done on the ways with the help of which machines learn by themselves.

ML process first need to input the training dataset in a particular algorithm. The training data trains the ML algorithm with known and unknown data [4]. Now to check that the trained algorithm is working properly, the algorithm is exposed to new input data. Then the results and predictions are checked. If the results are not as per expectations, then the algorithm has to train multiple times till it meet the desired result. This will enable the algorithm to learn continuously by its own and better result, which will increase accuracy percentage in output over time [5–7]. Today, both personal and professional lives are totally dependent on technology. Google assistant and Siri are the two examples. This is all because of ML and artificial intelligence [8–10].

11.2 Types of ML Algorithms

ML has various algorithms to train machines so that they can solve a problem. Based on the approach, it can be decided that which algorithm can be used. The different means by which a machine can learn and analyze the data are supervised learning (SL), unsupervised learning (UL), and reinforcement learning (RL) [11]. Figure 11.1 elaborates the different types of ML algorithms.

11.2.1 Supervised Learning

SL methods require external assistance. In this type of learning, external supervision is provided to a certain activity so that it can be done correctly. With the help of training dataset's input and responses, the SL algorithms make predictions for result of new datasets [12]. The way the machine is trained is known as supervised learning. The machine is provided with

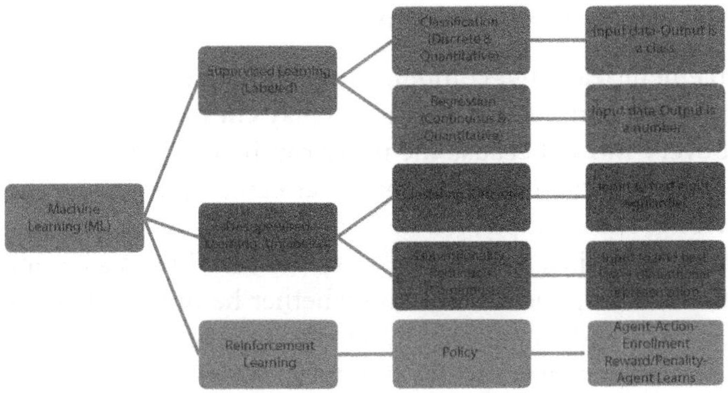

Figure 11.1 Types of ML algorithms.

input and corresponding answers are given. The training and test datasets are given to the machine as input. The algorithm learns different types of patterns from train dataset then analyze and do prediction by applying these to test datasets.

For example, if it is to be checked that what are the parameters of raining today, the humidity and temperature should be above certain level. The wind should be in a certain direction, so if this scenario is there, it will rain. Similarly, to let understand the kids with scenario, we tell them answers and example [13–15]. If the data is structured and can be classified on some basis, then SL can be applied on it.

11.2.2 Unsupervised Learning

In case of UL, the methods learn various features by the given data. The unsupervised methods use previously learned features to identify the classification of the data when new data is introduced. Unsupervised learning is mainly used for association and clustering. For example, when a kid is taking decisions out of their own understanding or through book, etc., this type of learning would be unsupervised learning. Here the computer is only given with the inputs and computer finds the pattern or structure in it. If the computer is given with inputs regarding fruits like what is the size, color, taste but the computer is not given the name of the fruit. Then, computer groups the fruits based on given characteristics finally comes out with the output [16, 17]. When the correlation in the data or structure of the data is not known, like in case of big data, which is huge chunk of unstructured data, unsupervised learning is used to find the structure. So, it is the job of the algorithm to find the structure, on behalf of which some decision can be made [18].

11.2.3 Reinforcement Learning

In reinforcement learning, computer tries to take decisions of their own. For example, if a computer is to train to play chess, then it is not possible to train it every move, because the move can be randomly changed in the game, so what one can do is, the computer can be told that is the move is right or wrong. For example, if a new situation comes up, the kid will take actions on his own, i.e., from his past experiences, but as a parent towards the end of an action, one can tell him whether he did good or not. In that case, the kid will understand that he should do repeat the action next time for same type of scenario or not. In a temperature control system, it has to decide whether to increase or decrease the temperature [19]. So, using reinforcement learning, using different parameters like number of persons in the room, outside temperature, etc., it makes decision with its past experiences. In this type, the hit and trail concept is used where the only way to learn is past experience. Table 11.1 describes about the variation between ML techniques based on various perspectives.

11.3 Applications of Machine Learning Techniques

11.3.1 Personal Assistants

As shown in Figure 11.2, Google, Bixby, Alexa, and Siri are some virtual personal assistants. Using neural language processing based algorithm, they help in searching information when asked. After activating, they can be asked for any type of information, setting schedule, calling on a number and sending commands to other phone apps for completion of the tasks. ML plays a significant role in collecting and refining information on the basis of previous experience with user [8].

11.3.2 Predictions

GPS navigation service is used all over the world. Whenever this app is used, the central server saves our current locations with velocities to manage a map of current traffic. This helps in estimating congestion on the basis of daily traffic experience. Accordingly, one can set the route. Also, cab booking app estimates ride's price and timing with the help of ML. Figure 11.3 shows few apps used for predictions [9].

Table 11.1 Difference between SL, UL, and RL.

	Supervised learning	**Unsupervised learning**	**Reinforcement learning**
Introduction	In this, external supervision is provided with the help of training data, to a certain activity so that it can be done correctly.	The unsupervised methods use previously learned features to identify the classification of the data when new data is introduced.	In reinforcement learning, computer tries to take decisions of their own.
Deals with problems related to	Regression problems and classification problems.	Problems which require clusters and problems related to anomaly detection.	The problems using hit and trail concept where the only way to take decision is the experience.
Required data type	Labeled data	Unlabeled data	No predefined data
Training requirements	Need external Supervision	No external supervision is required	No external supervision is required
Aim	Forecast Outcome	Discover underlying patterns	Understand a sequence of actions
Approach	Map labeled input to known output	Understand patterns and discover output	Follow trial and error method
Algorithms Names	Linear Regression, Support Vector Machine, Random Forest	C-Means, K-Means, a priori	SARSA, Q-Learning
Applications	Forecast Sales, Risk Evaluation	Anomaly Detection, Recommendation System	Gaming, Self-driving cars

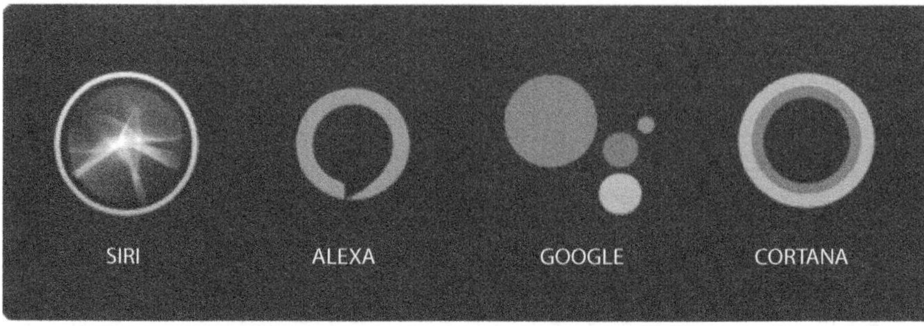

Figure 11.2 Personal assistants [8].

Figure 11.3 Apps used for navigations and cab booking [9].

11.3.3 Social Media

Social media utilizes machine learning for user and their own benefit. By understanding from experience, Facebook notices your connection with people, interests, profiles you often visit etc. then it suggests you the people who can be your friends [9]. So applications like face recognition and people you may know are very complicated at backend but at front end, these seems very simple application of ML [10]. Figure 11.4 is an example of using social media through mobile phone.

11.3.4 Fraud Detection

Fraud detection is an important and necessary application of ML. The number of frauds are increasing day by day due to more payment channels like numerous wallets, credit/debit cards tec. Also, the number of criminals have become proficient at searching loopholes. When a person performs some transaction, the ML method search profile for suspicious patterns. These kinds of problems are classification problems in machine learning [10].

Figure 11.4 Social media using phone [10].

Figure 11.5 Fraud detection [10].

Figure 11.6 Google translator [10].

11.3.5 Google Translator

Gone are the days when it was difficult to communicate in areas having other than native language. Figure 11.6 show icon of Google translator. Google's Neural Machine Translation is a machine learning translator which uses Natural Language processing and works on various languages and dictionaries. This ML application is mostly used application [10].

11.3.6 Product Recommendations

Online shopping websites recommends items those somehow matches with customer's taste. Websites or apps are able to do so using ML. Based on past experience of site visiting, product selection, brand preferences etc., the product recommendation is done [9, 10] (refer Figure 11.7).

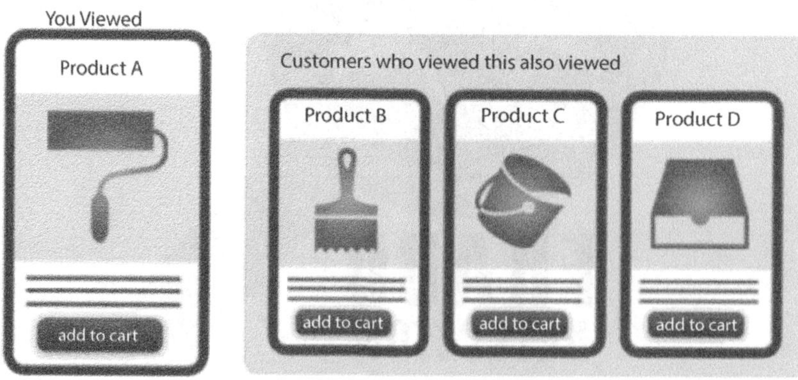

Figure 11.7 Product recommendations [9].

Figure 11.8 Surveillance with video [10].

11.3.7 Videos Surveillance

It is quite difficult for a single person to monitor multiple video cameras. So, computers are trained to make this job easy. Video Surveillance is an application of artificial intelligence that detect crime before happening. By tracking unusual activities, like stumbling, meaningless standing of someone for a long time etc., the system alerts the human attendants to avoid mishaps. This task is performed actually with the help of ML at backend [10] (refer Figure 11.8).

11.4 Solution to a Problem Using ML

The data science problems can be categories in five ways which can be understood by following five questions given in the diagram.

11.4.1 Classification Algorithms

These types of algorithms classify a record. We can use these for a question with limited count of answers. If the problem wants an answer of first type of question in Figure 11.9, for example, "Is it cold," then classification algorithms are used. It works for questions having certain number of answers like true/false, yes/no, or maybe. The first question in the diagram has two choices, so it is called two-class classification, and if the question has more than two choices then it is called multiclass classification [20].

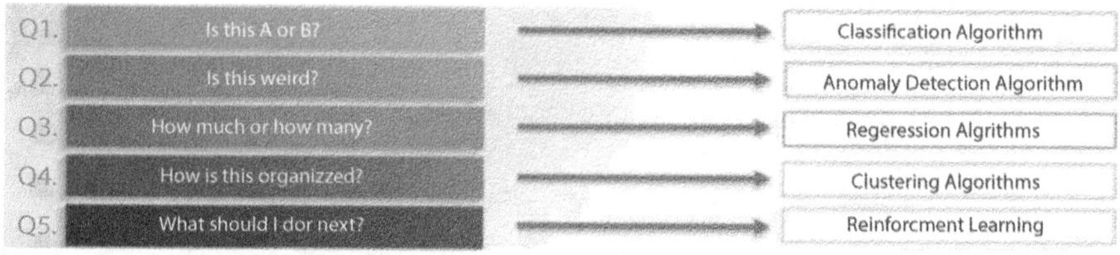

Figure 11.9 Data science problem categories [20].

11.4.2 Anomaly Detection Algorithm

This type of algorithm alerts for change in some particular pattern when analyze it. So, if the problem is to analyze unusual happening and where one wants to find anomaly or odd one out, then Anomaly Detection Algorithms are used.

In Figure 11.10, there is a pattern of all blue persons, but when one red man comes in between, which can be called as anomaly, the algorithm will flag that person because he was someone who was not expected [21]. In real life, credit card companies use these anomaly detection algorithms to flag any transaction, which is not usual as per the company's transaction history and put message on the registered number to confirm that the transaction is done by authenticated person.

11.4.3 Regression Algorithm

Regression analysis investigates relationship in an independent variable(s) and dependent variable. Regression algorithms can be used to calculate a continuous value such as weight or salary. These algorithms fall in supervised learning category. These are used to calculate numeric values using formulas. In these types of algorithms, we deal with questions like "what should be the number of hours one should put in to

Figure 11.10 Anomaly detection in red color person [21].

get promotion?" i.e., the problems where we want a numeric value [12]. There are different models with regression analysis. The most important among all regression-based algorithms are linear and logistic regressions.

11.4.4 Clustering Algorithms

Clustering algorithm helps to understand the structure of a dataset. These algorithms separate the data into groups or clusters, to ease out the interpretation of the data. Data organization helps in prediction of behavior of some event. So, when the structure behind a dataset is to find out, then clustering algorithms are used [21] (refer Figure 11.11).

In unsupervised learning, where one tries to establish a structure from unstructured data, clustering algorithms are used. If one feeds data to computer, then applies clustering algorithm on it, it categories the data into groups A, B, C on behalf of which one can make decision that what he can do with this data.

11.4.5 Reinforcement Algorithms

This type of algorithms deal with the problems where lots of inputs given to machine and we want to take some decision on the basis of past experiences. These algorithms were designed as to how brains of responds to punishments and reward, they learn from past results and then decide on next action. They are good for the systems, which require small decision making without human assistance.

These algorithms analyze the dataset using trial and error method and predict the output with higher rewards. The three main components used in reinforcement learning are the agent, environment and actions. Here the

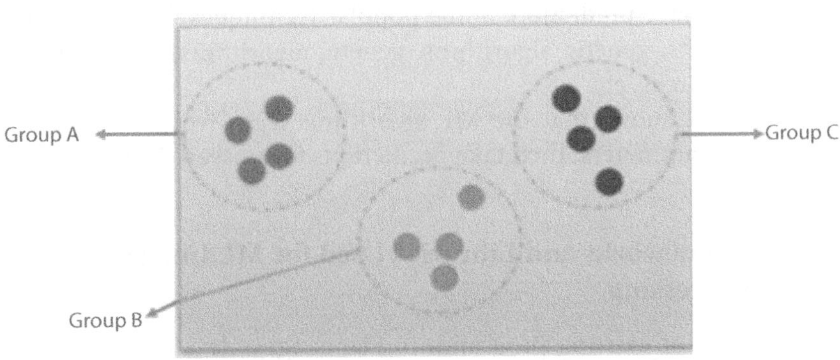

Figure 11.11 Data clustering [21].

agent is a learning machine, the environment means the conditions with which the agent interacts and finally with past experience and predicted data, the agent makes a decision and performs certain action [19]. Table 11.1 summarizes the difference between the 3 types of ML techniques on the basis of different criteria.

11.5 ML in Image Processing

Computer vision is a field where machines can recognize videos and images. The core of this field is image processing. Image processing is a technology that can process the images, analyze them, and can extract the meaning details from these images. This field is used now a days in several areas for various purposes like pattern recognition, visualization, segmentation, classification etc. Image processing can be applied using two methods- analogue image processing and digital image processing. The former method is used for hard copy images. For example- scanning printouts. The latter is used to manipulate the digital images to extract meaningful information about them. ML and deep learning-based techniques are becoming more popular for image processing. These techniques interpret images like human brain. Some examples of image processing using ML are biometric authentication, gaming with virtual reality experience, image sharpening, self-driving technology, etc. Images are to be processed to be more suitable for using them as input. For example, images are to be converted from PNG or JPEG to byte data or array of pixels form for neural networks. So here, computer vision term is to generate ideal data sets for ML techniques after processing and manipulating images. For example, to predict an image is of a cat or a dog. For this, collection of cat and dog images is made and processed to extract features to be used by the ML techniques to have prediction. Some popular techniques for this purpose are—neural nets, genetic algorithms, genetic algorithms, nearest neighbors, decision trees, etc.

Figure 11.12 shows that the ML algorithms learn from training data with specific parameters, then take predictions for unseen data.

11.5.1 Frameworks and Libraries Used for ML Image Processing

Among plenty of existing programming languages, developer preferably use python for ML applications. However, other languages can also be used

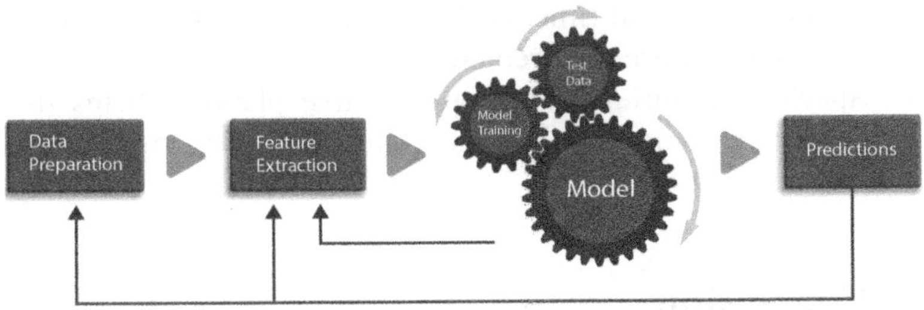

Figure 11.12 Workflow of image processing using ML data clustering [22].

which are suitable for particular use case. The frameworks used for various ML image processing applications are [22]:

- OpenCV—This is python library which is used in solving many computer-vision problems. This is open-source framework that works with both videos and images.
- Tensorflow—It is a framework developed by Google that is very popular for ML applications. It is also one open-source framework that provides huge library of ML algorithms. It works for cross platform too.
- PyTorch—This is developed by Facebook and is very popular for neural network applications. It implements distribution training, provides cloud support and again is an open-source framework.
- Caffe—This framework is very popular deep learning-based framework that provides modularity and speed. This Berkeley AI Research developed framework is based on C++ language and has expressive architecture.
- EmuguCV—This framework works with all languages compatible with .NET. It also works for cross-platform.
- MATLAB toolbox for image processing—This toolbox consists of huge library of image processing techniques based on deep learning and interactive image processing 3D workflows, also helps in automate these. One can apply segmentation on the datasets, process large datasets batch wise and can perform comparison between different image registration methods.
- WebGazer—This framework consists of a huge library that is used for eye tracking. Using standard webcams, this provides information of eye-gaze sites to the web visitors while

surfing web in real-time. With this, one does not require any specific hardware requirements.
- Apache Marvin AI—This open-source platform helps in delivering complex solutions while simplifying modelling and exploitation.

11.6 Conclusion

ML is a subclass of AI and is one of the most powerful technology now. It is a tool to turn the information into knowledge. The ample data produced in last 50 years is useless till we analyze and find hidden patterns from it. ML uses data and results to predict the rules behind a problem. This paper gives an overview of ML basics, types of algorithms and applications. This paper includes some open source libraries which are utilized for preprocessing, analyzing, and extracting the details from the images with the help of ML. Although the paper is not resolving this substantial concept, hopefully it clears the basic concepts and provides useful information.

References

1. Kotsiantis, S.B., Zaharakis, I.D., Pintelas, P.E., Machine learning: A review of classification and combining techniques. *Artif. Intell. Rev.*, 26, 3, 159–190, 2006.
2. Kato, N., Mao, B., Tang, F., Kawamoto, Y., Liu, J., Ten challenges in advancing machine learning technologies toward 6G. *IEEE Wirel. Commun.*, 27, 3, 96–103, 2020.
3. Mehrabi, N., Morstatter, F., Saxena, N., Lerman, K., Galstyan, A., A survey on bias and fairness in machine learning. *ACM Comput. Surv. (CSUR)*, 54, 6, 1–35, 2021.
4. Bengio, Y., Learning deep architectures for AI. *Found. Trends Mach. Learn.*, 2, 1–127, 2009.
5. Dhall, D., Kaur, R., Juneja, M., Machine learning: A review of the algorithms and its applications. *Proceedings of ICRIC 2019*, pp. 47–63, 2020.
6. Dietterich, T.G., Machine learning for sequential data: A review, in: *Joint IAPR International Workshops on Statistical Techniques in Pattern Recognition (SPR) and Structural and Syntactic Pattern Recognition (SSPR)*, pp. 15–30, Springer, Berlin, Heidelberg, 2002.
7. Rogers, S. and Girolami, M., *A first course in machine learning*, Chapman and Hall/CRC, 2016. https://doi.org/10.1201/9781315382159.

8. Hassanien, A.E., Tolba, M., Taher Azar, A., Advanced machine learning technologies and applications, in: *Second International Conference,* Egypt, AML, Springer, 2014.
9. https://medium.com/app-affairs/9-applications-of-machine-learning-from-day-to-day-life-112a47a429d0
10. https://www.edureka.co/blog/machine-learning-applications/
11. Machine learning algorithms: A review. *Int. J. Comput. Sci. Inform. Technol.*, 7, 3, 1174–1179, 2016.
12. Singh, A., Thakur, N., Sharma, A., A review of supervised machine learning algorithms, in: *2016 3rd International Conference on Computing for Sustainable Global Development (INDIACom)*, IEEE, pp. 1310–1315, 2016.
13. Kotsiantis, S.B., Zaharakis, I., Pintelas, P., Supervised machine learning: A review of classification techniques, in: *Emerging Artificial Intelligence Applications in Computer Engineering,* vol. 160, pp. 3–24, 2007.
14. Choudhary, R. and Kumar Gianey, H., Comprehensive review on supervised machine learning algorithms, in: *International Conference on Machine Learning and Data Science (MLDS)*, IEEE, pp. 37–43, 2017.
15. M.A.R. Schmidtler and R. Borrey, Data classification methods using machine learning techniques. U.S. Patent 7,937,345, May 3, 2011.
16. Ball, G.R. and Srihari, S.N., Semi-supervised learning for handwriting recognition." Document analysis and recognition. *2009. ICDAR'09. 10th International Conference on*, IEEE, 2009.
17. Sharma, D. and Kumar, N., A review on machine learning algorithms, tasks and applications. *Int. J. Adv. Res. Comput. Eng. Technol. (IJARCET)*, 6, 10, 1548–1552, 2017.
18. Al-Hmouz, A., Shen, J., Yan, J., A machine learning based framework for adaptive mobile learning, in: *Advances in Web Based Learning–ICWL 2009*, pp. 34–43, Springer Berlin Heidelberg, 2009. http://dx.doi.org/10.1007/978-3-642-03426-8_4.
19. Szepesvári, C., Algorithms for reinforcement learning, in: *Synthesis lLectures on Artificial Intelligence and Machine Learning*, vol. 4, pp. 1–103, 2010.
20. Kotsiantis, S.B., Zaharakis, I., Pintelas, P., Supervised machine learning: A review of classification techniques, in: *Emerging Artificial Intelligence Applications in Computer Engineering*, vol. 160, pp. 3–24, 2007.
21. Shon, T. and Moon, J., A hybrid machine learning approach to network anomaly detection. *Inf. Sci.*, 17718, 3799–3821, 2007.
22. https://nanonets.com/blog/machine-learning-image-processing/ [Date: 11/11/2021]

12

Use and Application of Artificial Intelligence in Accounting and Finance: Benefits and Challenges

Ram Singh[1]*, Rohit Bansal[2] and Niranjanamurthy M.[3]

[1]Quantum School of Business, Quantum University Roorkee, India
[2]Department of Management, Vaish Engineering College Rohtak, India
[3]Department of AI and ML, BMS Institute of Technology and Management, Bangalore, India

Abstract

Background and Introduction: AI is significant in Accounting and Finance as it smoothens out and improves numerous tedious bookkeeping measures. The general result is that associations can set aside additional time and cash as AI gives significant bits of knowledge to bookkeeping and monetary investigators and helps with dissecting a lot of information quick, producing more precise, noteworthy information at lower costs. This information would then be able to be utilized to convey bits of knowledge and examination, driving key dynamic that influences the entire organization.

Purpose and Method: The main objective of the chapter is to find out the use and application of Artificial Intelligence in the sector of Accounting and Finance. The idea of the examination is engaging, which depends on auxiliary information and data. The necessary information and data have been gathered from different sites, diaries, magazines, and media reports.

Discussion and Conclusion: Finally, the segment reasons that AI machines guarantee useful capability while restricting prizes. As computerization is getting to each edge of the business, the financial firm will moreover accept the high-level change that achieves from the advancement enhancements. The accounting and finance that passed on AI will be situated in the destiny of electronic changes; there are different advantages in accounting due to man-made intellectual prowess.

**Corresponding author*: ramsinghcommerce@gmail.com

M. Niranjanamurthy, Kavita Sheoran, Geetika Dhand, and Prabhjot Kaur (eds.) Data Wrangling: Concepts, Applications and Tools, (251–274) © 2023 Scrivener Publishing LLC

Keywords: Artificial intelligence, machine learning, NLP, chatbots, robotic process automation

12.1 Introduction

The articulation "Artificial Intelligence" was organized at a gathering at "Dartmouth College in 1956." Until 1974, AI included work that idea for dealing with issues for math and variable-based math and giving in typical language. Some place in the scope of "1980 and 1987," there was a climb in ace systems that tended to requests or handled issues about unequivocal data. "Interest in AI declined until IBM's Deep Blue, a chess-playing PC, crushed Russian grandmaster Garry Kasparov in 1997," from that point forward, other AI accomplishments have come to incorporate penmanship acknowledgment, testing for self-ruling vehicles, the primary homegrown or pet robot, and humanoid robots. "Artificial Intelligence" has effectively reformed various businesses, including medical services. It is acquiring speed and we can observer numerous advancements that appeared to be unthinkable a couple of years prior. "Each tech merchant and science organization associated with clinical exploration or clinical preliminaries endeavors to make dependable prescient and prescriptive instruments for both diagnosing and treatment, the innovation research organization Gartner accepts that 75% of medical care associations will have put resources into their AI potential by 2021 to improve the general exhibition [18], the upsides of AI-driven clinical devices are important and advantageous for specialists, and patients, and are appropriate in various medical services regions." AI has numerous applications in a bunch of enterprises, including money, transportation, and medical care which will change how the business judgments and treats ailments. "Artificial Intelligence" has been applied to protest, face, discourse, and penmanship acknowledgment; computer generated reality and picture handling; characteristic language preparing, chatbots, and interpretation; email spam separating, advanced mechanics, and information mining. As per market knowledge firm, Tractica, the yearly overall AI income will develop to "$36.8 billion by 2025."

12.1.1 Artificial Intelligence in Accounting and Finance Sector

"AI suggests the generation of human understanding in machines that are adjusted to think like individuals and duplicate their exercises, the term may in like manner be applied to any machine that shows characteristics identified with a human mind, like learning and basic reasoning." The ideal property of man-made thinking is its ability to pardon and take actions that have clear chance concern achieving a specific target and a "subset of man-made thinking is AI," which insinuates the possibility that PC activities can thusly acquire from and conform to new data "without being helped by individuals." "Significant learning procedures enable this customized learning by ingestion of immense unstructured data estimates like substance, pictures, or video." Computerized reasoning is a part of PC sciences that stresses the headway of clever machines, thinking and performing errands very much like people. A portion of the principle utilizations of Artificial Intelligence incorporate discourse acknowledgment (Figure 12.1), Natural language processing (NLP), machine vision, and expert frameworks, AI is assuming an imperative part in the computerized change of bookkeeping and money [11]. Computer-based intelligence machines will assume control over the weight of doing tedious and tedious assignments, AI in bookkeeping diminishes human mediation and AI applications and AI administrations assist with financing specialists achieving their normal obligations quicker [1, 2].

By and large, the job of money experts is to make methodologies to convey business resources. While, accountants job is to record and report each monetary exchange of business, blunders while recording

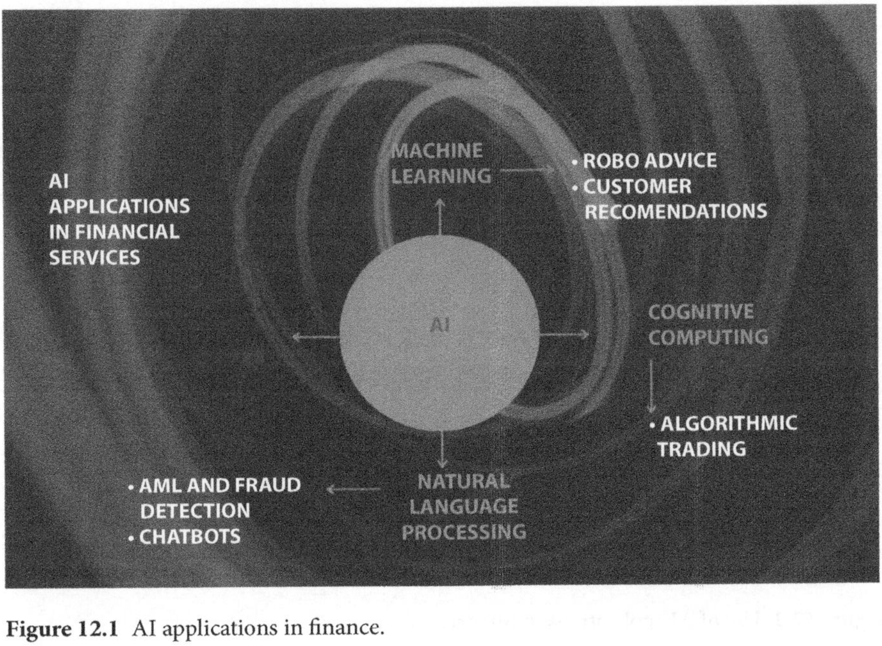

Figure 12.1 AI applications in finance.

monetary exchanges, review mix-ups, and obtainment measure mistakes are the recent concerns that bookkeeping experts are confronting today. Simulated intelligence advances like machine learning (ML) in bookkeeping and profound learning helps bookkeeping and finance to play out their undertakings all the more effectively. With this, we can comprehend that AI upholds the human labor force, however does not take their positions. Thus, the advanced change in the bookkeeping and money area utilizing AI is inconceivable. Computer-based intelligence bookkeeping programming carries an uncommon change to your business, few chances that this most progressive man-made reasoning programming assists with digitizing the money and bookkeeping assignments totally [3–5].

12.2 Uses of AI in Accounting & Finance Sector

12.2.1 Pay and Receive Processing

Existing AI-based receipt the leaders' structures are helping finance clients in making receipt planning viably. Progressed change in accounting and cash is incomprehensible and progressed machines using AI (Figure 12.2) are learning the accounting codes that best suits each receipt.

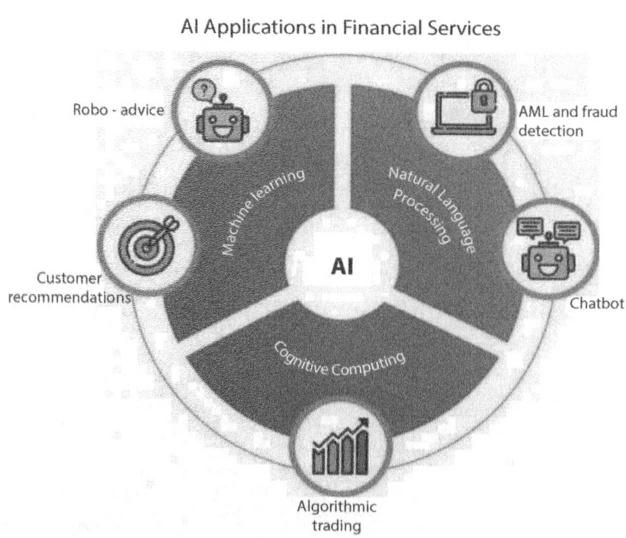

Figure 12.2 Use of AI applications in finance.

12.2.2 Supplier on Boarding and Procurement

The AI-based structures can screen the suppliers by taking a gander at their evaluation nuances or FICO appraisals. Artificial intelligence devices can set all suppliers in the systems without the prerequisite for individuals; similarly, they can moreover set the request sections to get the fundamental data. "Numerous associations archive their procurement and buying methods on paper, they stay aware of various systems and records that are not in regards to each other, and as AI machines measure unstructured data using APIs, the acquisition communication will be robotized" [6, 7].

12.2.3 Audits

Digitization in the review cycle upgrades the security level (Figure 12.2). Utilizing an advanced tracker, the reviewers can follow each record that is gotten too. Maybe than looking through all paper records, computerized documents can facilitate the review work. Henceforth, the digitization cycle in inspecting gives further developed exactness of reviews, consequently, computerized reasoning in bookkeeping and reviewing assists with recording each monetary exchange of the organization, Artificial intelligence controlled reviews are more productive and great [8, 9].

12.2.4 Monthly, Quarterly Cash Flows, and Expense Management

Computer-based intelligence controlled machines can assemble data from many sources and organize that data. Computerized reasoning devices, contraptions, or AI applications speed up cycles, yet they make money related cycles careful and secure, month to month, quarterly, or yearly salaries will be accumulated and consolidated adequately by AI controlled machines. Changing and finishing up expenses to attest that they are pleasing according to the association's norms is an irksome task (Figure 12.2). "The manual cycle eats up more freedoms for your cash bunch. Maybe than individuals, machines can do these tasks quickly and viably, AI machines can examine all receipts, survey the costs, and moreover alert the human workforce individuals when a break occurred" [10].

12.2.5 AI Chatbots

AI driven Chatbots are created to settle client's inquiries productively (Figure 12.2). The questions might incorporate the most recent record balance subtleties, explanations, credit bills, and record status, and so on In this manner, AI for bookkeepers is helping from various perspectives and USM AI administrations and answers for bookkeeping and money can accomplish for your business, everyday advances in AI innovation is taking bookkeeping to the most significant levels [12].

12.3 Applications of AI in Accounting and Finance Sector

"AI might perhaps change the cash and accounting undertakings with types of progress that crash dreary tasks and free human cash specialists to accomplish more raised level and more beneficial assessment and coordinating for their clients." Be that as it may, affiliations keep thinking about whether to use AI in their workforce due to weaknesses around the business case or benefit from hypothesis, AI has been executed in a couple of adventures from stock trading to facilities. "Google has singled it out as the accompanying gigantic thing, one of the chief difficulties for the clerks is the colossal proportion of trades that the customers may have to oversee especially in the B2B space where you have hundreds and thousands of customers and an enormous number of sales and you need to seek after each trade." So that is where a huge load of time is being spent by having bunches actually oversee gigantic trades. So when you need to follow such endless trades, following each trade, there comes the work of development. Hence, finance gatherings' post for Business Accounting Software and mechanical assemblies to restrict common contingent activities, allowing them to redirect their accentuation on examining data, giving critical arrangement, and truly advancing the business. "Forbes predicts that by 2020, accounting tasks including charge, money, surveys, and banking will be totally motorized using AI-based advances, which will agitate the Accounting Industry in habits never imagined and bring both tremendous opportunities and certifiable hardships." Simulated intelligence pledges to help both effectiveness and nature of yields while permitting more vital straightforwardness and survey limit. Not simply, AI will give a sweeping extent of possible results and breaking point the standard commitments of the cash bunch anyway it will moreover save time and allow accounting specialists an opportunity to coordinate basic assessment on alternate points of view. Other than that, AI will adequately guess precise spending outlines. The focal thought is that with AI, accounting specialists would expect future data subject to past data, with key business benefits and squeezing factors from all around educated customers top of the mind,

AI computations are being done by FIs across each money-related assistance here is the mystery:

12.3.1 AI in Personal Finance

Buyers are eager for monetary autonomy, and giving the capacity to deal with one's monetary wellbeing is the main thrust behind the reception of AI in individual budget. Regardless of whether offering day in and day out monetary direction through chatbots fuelled by regular language handling or customizing experiences for abundance the executives arrangements, AI is a need for any monetary establishment seeming to be a top part in the business. An early illustration of AI in individual budget is Capital one's Eno. "Eno dispatched in 2017 and was the primary normal language SMS text-based right hand offered by a US bank. Eno produces experiences and expects client needs through more than 12 proactive capacities, for example, cautioning clients about presumed misrepresentation or value climbs in membership administrations" [16].

12.3.2 AI in Consumer Finance

Artificial intelligence can examine and single out inconsistencies in designs that would some way or another go unrecognized by people. One bank exploiting AI in purchaser finance is JPMorgan Chase. For Chase, purchaser banking addresses more than half of its net gain; all things considered, the bank has embraced key extortion distinguishing applications for its record holders. For instance, it has carried out an exclusive calculation to identify misrepresentation designs each time a Visa exchange is handled, subtleties of the exchange are shipped off focal PCs in Chase's server farms, which then, at that point choose whether or not the exchange is false. "Pursue's high scores in both Security and Reliability to a great extent supported by its utilization of AI procured it second spot in Insider Intelligence's 2020 US Banking Digital Trust study" [19].

12.3.3 AI in Corporate Finance

"AI is particularly helpful in corporate cash as it can all the more promptly expect and review credit possibilities." For associations expecting to fabricate their value, AI progresses, for instance, AI can help with additional creating credit ensuring and decrease the financial risk. "AI can moreover diminish money related bad behavior through state of the art coercion disclosure and spot strange development as association clerks, specialists, lenders, and monetary patrons pursue long stretch turn of events." U.S. Bank is using AI in the two its middle and managerial focus applications and they opens and takes apart terrifically significant data on customers through significant sorting out some way to help with recognizing agitators; it has been using this development against illicit duty evasion and, according to an Insider Intelligence report, has increased the yield differentiated and the previous structures' ordinary limits [20].

12.4 Benefits and Advantages of AI in Accounting and Finance

"AI Chatbots, Machine Learning Tools, Automation, and other AI headways are expecting a major part in the cash region, Accounting and Finance affiliations have been placing assets into these progressions and making them a piece of their business." New development is changing the way in which people work in every industry. It is similarly changing the suspicions clients have when working with associations and AI can help accountants with being valuable and useful, and 80% to 90% decline in the time it takes to complete tasks will allow human accountants to be more focused on offering direction to their clients. Adding AI to accounting undertakings will similarly construct the quality since botches will be diminished. When accounting firms embrace man-made thinking to their preparation, the firm ends up being more engaging as a business and expert center to twenty to thirty-year-olds and Gen Z specialists. "This partner grew up with development, and they will expect that forthcoming bosses ought to have the latest advancement and headway to help not simply their working tendencies of versatile schedules and far off regions yet, what is more, to let free them from customary tasks that machines are more able to wrap up." As clients, twenty to thirty-year-olds and Gen Zers will sort out whom to work with relying upon the help commitments they can give. As extra accounting firms take on AI prowess, they will really need to give the data encounters made possible through computerization while the people who do not zero in on the development cannot fight. "Robotic Process Automation (RPA)" grants machines or AI workers to complete repetitive, dreary endeavors in business cycles, for instance, file examination and dealing with that are plentiful in accounting, when RPA is set up, time clerks used to spend on these tasks is as of now open for more key and cautioning work [22].

AI can imitate human coordinated effort all around, for instance, understanding actuated importance in client correspondence and using bona fide data to acclimate to an activity. "AI often give the continuous status of a financial issue since it can manage chronicles using typical language getting ready and PC vision faster than any time in ongoing memory making step by step enumerating possible and unassuming" [23]. This information licenses associations to be proactive and shift direction if the data show negative examples, the electronic endorsement and planning of records with AI advancement will further develop a couple of internal accounting measures including procurement and purchasing, invoicing, purchase orders, cost reports, lender liabilities, and receivables, and that is only the start. In accounting, there are various internal corporate, closes by, state and government rules that ought to be followed. "AI engaged structures help support examining and ensure consistence by having the choice to screen chronicles in opposition to rules and laws and flag those with issues, and deception costs associations everything considered billions of dollars consistently and money related organizations associations have $2.92 in costs for every dollar of blackmail" [18]. AI estimations can quickly channel through immense proportions of data to see potential deception issues or questionable development that might have been by and large missed by individuals and pennant it for extra thought.

12.4.1 Changing the Human Mindset

It seems like the singular limit to AI mental ability gathering in accounting is getting people lively with regards to the change, practically 85% of pioneers grasp that AI will help their associations with accomplishing or backing a high ground. "The CEOs seem to appreciate the meaning of Artificial Intelligence; it basically requires a viewpoint shift from the accounting specialists to recognize the changes, and with an assistance from AI-engaged systems, clerks are opened up to gather relationship with their clients and pass on fundamental encounters." To help accountants with enduring and in a perfect world hug the tech development to accounting firms, it is basic that the upsides of robotization and Artificial Intelligence are conferred to them and they are outfitted with the fitting readiness and any assistance critical to sort out how best to use AI for their likely advantage. "AI and motorization in accounting and cash are just beginning, regardless, the development is getting more perplexing, and the mechanical assemblies and structures available to help accounting are developing at a quick speed" [13]. Accountants that go against these movements cannot keep up with up with others who partake in the advantage of time and cost venture assets and encounters AI can give.

12.4.2 Machines Imitate the Human Brain

"Robotization, AI chatbots, AI gadgets, and other AI propels are accepting a principle part in the cash region, accounting and cash associations are making them a piece of their business by putting strongly in these progressions" [14]. As demonstrated by examiners, AI applications and ML applications are influencing accounting and cash specialists and their normal positions, using AI and ML, finance experts can additionally foster convenience and oversee new customers. "AI can displace individuals with the monotonous control of eliminating, assembling, and arranging the data, in any case, those identical clerks and analysts working with AI can perform different tasks" [6]. Regardless, they show the AI what data to look for and how to figure out it. Then they look at irregularities. Thusly, AI can take on the somewhat long mix that possesses such a great deal of time for data segment and think twice about moreover forgo botches, reducing commitment with the common tasks dealt with, and clerks will be permitted to partake in additional notice occupations.

12.4.3 Fighting Misrepresentation

"With the help of AI computations, portions associations can analyze more data in new and imaginative habits to recognize any manufacture development, and every purchaser trade joins especially unmistakable information." With AI, and AI portions associations can look rapidly and capably through this data past the standard course of action of components like time, speed, and total [21]. "Computer-based intelligence helps in capably planning monstrous proportions of data from different sources, pay extraordinary psyche to precarious trades and associations, and report them in a visual instrument that, hence, will allow the consistence gathering to manage such sorts of questionable cases even more feasibly."

12.4.4 AI Machines Make Accounting Tasks Easier

As per a counselling firm Accenture, "Robotization, minibots, AI, and versatile insight in the wake of turning into a piece of the money group at lightning speed." AI machines computerize bookkeeping methodology all over, it guarantees functional productivity while decreasing expenses. "As computerization is getting to each edge of an association, the financial associations moreover embrace the high level change that will gain from the development enhancements and the accounting and cash pioneers who passed on AI will be situated in the destiny of mechanized changes" [17].

For example, Xero, an accounting firm, has dispatched the Find and Recode computation that modernizes the work and finds typical models by separating code corrections. Using the computation, 90% more definite results were found while separating 50 sales.

12.4.5 Invisible Accounting

AI considers dull errands to be wiped out from a representative's everyday responsibility, and furthermore builds the measure of promptly accessible information readily available. This, thusly, expands the insight accessible to comprehend the wellbeing and course of a business at some random time. Simulated intelligence consequently deals with the way toward social event, arranging, and envisioning appropriate information such that helps the business run all the more productively. This opens up staff to accomplish more useful errands and gives them more opportunity to drive the business advances.

12.4.6 Build Trust through Better Financial Protection and Control

AI can likewise fundamentally diminish monetary misrepresentation and limit bookkeeping blunders, frequently brought about by human oversight. The ascent of web-based banking has brought a large group of benefits; however it has additionally made new roads for monetary wrongdoing, explicitly around extortion. The odds of an unscrupulous installment falling through the net develop as the volumes of information increment. "That has made the bookkeeper's consistence task a lot harder to finish and AI can deal with that information audit at speed." It can likewise assist with appointing costs to the right classes, guaranteeing the organization does not pay out for things it should not, by executing mechanized enemy of misrepresentation and money the executive's frameworks, practices can altogether further develop consistence strategies and ensure both their own and customers' accounts [13]. Thusly, AI and bookkeepers can cooperate to give a more prescient, vital assistance utilizing the accessible information to get on expected issues before they emerge.

12.4.7 Active Insights Help Drive Better Decisions

Notwithstanding the area, AI can be utilized to break down enormous amounts of information at speed and at scale. It can distinguish inconsistencies in the framework and enhance work process. Money experts can utilize AI to help with business dynamic, in light of noteworthy experiences got from client socioeconomics, past conditional information, and outside factors, all progressively. It will empower bookkeepers to think back as well as look advances with more lucidity than any time in recent memory, and organizations can utilize information to perform income estimating, anticipating when the business may run out of cash, and make moves to ensure against the circumstance early [15]. They can recognize when a client may be going to beat and see how to restore their series, this whole method is that bookkeepers will actually want to assist customers with reacting monetary difficulties before they become intense, changing consumption or cycles as required. "As AI coordinates more extensive business data streams into the bookkeeping blend, bookkeepers can likewise widen their prescient consultancy past unadulterated monetary wanting to join different spaces of the business."

12.4.8 Fraud Protection, Auditing, and Compliance

Applying AI to informational collections can likewise help in diminishing misrepresentation by giving persistent monetary examining cycles to ensure organizations are in consistence with nearby, government, and, if relevant, global guidelines. Computer-based intelligence utilizes its calculations to quickly figure out enormous informational indexes and banner expected extortion and dubious movement. It comes through past practices of various exchanges to feature odd practices, for example, stores or withdrawals from different nations that are now and again bigger than ordinary aggregates. Simulated intelligence additionally ceaselessly gains from GL reviews and rectifications by people or hailed exchanges so it can improve decisions later on. Moreover, AI assists with decreasing extortion with advanced banking, particularly as the volume of exchanges and information increments. It searches for dubious and exploitative installments that might have escaped everyone's notice because of human blunder. Perhaps the most significant, yet dreary positions of bookkeeping groups are evaluating their information and records to be in consistence with unofficial laws. Man-made intelligence applies ceaseless GL or recordkeeping inspecting and catches business exercises and exchanges progressively. By performing nonstop compromises and acclimations to gatherings, an organization's books are more exact consistently, while eliminating a portion of the weights of month-end close for money and bookkeeping groups. Man-made intelligence empowered calculations in this product utilize these reviews to assist with guaranteeing the organization's reports and cycles are keeping the laws and rules set out by various government establishments. As can be found in the chart beneath, G2 information mirrors this hunger for programming to assist organizations with overseeing and mechanize their installment measures. This can be seen by the spike in rush hour gridlock to the "Enterprise Payments Software" and "AP Automation Software" classifications in March 2020 when the lockdown because of the COVID-19 pandemic started in the United States.

Nym Health raises $16.5 million for its auditable AI apparatuses for robotizing emergency clinic charging Nym, which has constructed a stage to computerize income cycle the board for medical clinic charging, has recently raised $16.5 million including subsidizing from Google's endeavor arm, GV. Their AI apparatuses assist medical clinics with the lasting issue of charging, which can be especially troublesome because of convoluted coding. Their product changes over clinical graphs and electronic clinical records from doctor's discussions into appropriate charging codes naturally. As indicated by Nym, "the organization utilizes normal language preparing and scientific classifications that were explicitly evolved to comprehend clinical language to decide the ideal charge for every methodology, assessment and symptomatic led for a patient." Billing difficulties are an especially troublesome issue inside the medical care space and across different businesses, like monetary administrations and retail. Innovation dependent on Natural Language Processing (NLP) can assist with financing divisions in any of these enterprises sort out bills, solicitations, and that is only the tip of the iceberg, and across classes on G2, organizations have been quick to digitize and mechanize their work and work processes for a significant length of time now. Because of the COVID-19 pandemic, organizations have run to G2 since March 2020, searching for approaches to work more intelligent. We saw a significant uptick in rush hour gridlock to the site, with organizations hoping to abbreviate timetables and develop rapidly at scale.

12.4.9 Machines as Financial Guardians

All enterprises require experts who act in both monetary and legitimate viewpoints and shoulder huge monetary onus. They are normally exceptionally talented and experienced workers, however being human; they are thusly inclined to confusions, predispositions, and other human mistakes. Machines and PCs with modern AI capacities can some time or another take over such monetary jobs as they are not inclined to such human blunders and become more viable guardians than their human partners. This part of AI execution is profoundly alluring for public trust subsidizes like clinical examination, parks, instructive organizations, and so forth, by guaranteeing long haul congruity and adherence to the first commands.

12.4.10 Intelligent Investments

AI upheld venture the board or "robotized abundance directors" as indicated by The Economist, are bound to offer sound monetary exhortation without bringing on board a full-time consultant. "The receptive discernment-based advantages of AI have provoked the curiosity of the worldwide venture local area too and Bridgewater Associates, one of the greatest multifaceted investments directors on the planet, have effectively evolved AI-supported exchanging calculations, which are equipped for foreseeing market patterns dependent on verifiable and factual information" [23]. "While such AI frameworks will consider more prominent upper hands for singular financial backers, it still anyway additionally represents an incredible danger to the market, in the event that each financial backer out there is equipped with such AI frameworks, it may have critical impeding impacts on the whole market as it will incredibly impact capital streams and macroeconomic approaches."

12.4.11 Consider the "Runaway Effect"

It is profoundly legitimate given the psychological capacities of AI frameworks, that they may, sooner or later, create self-governing information/information. Programming codes and calculations, initially intended to guarantee ideal framework productivity, could bring about adverse circumstances. This impact, known as the "Runaway impact" which causes the very things we tried to fix or tackle to go south on us and do significantly more noteworthy mischief [27]. With AI, the runaway impact, if at any point present will make a larger number of issues than it settles, and deciding by the current degree of AI refinement, the day where AI can be depended upon to moderate all adverse results is still very far away.

12.4.12 Artificial Control and Effective Fiduciaries

AI-based machines will actually want to take over many undertakings until recently connected to bookkeepers and HR work force. Most significant is the capacity to control the elements of legal consistence of different standards and guidelines. It can likewise assess worker execution which thusly can impact HR dynamic. Many believe this to be a startling part of intrusion of human security since the investigation of way of life examples and human conduct will be made by a "clever" machine. The inquiry then, at that point is can these machines find some kind of harmony between distinct information investigation and a more profound human-like sympathy while showing up at choices [23]. Accountants are people who play a vital part to play in any business and take significant obligations on monetary angles. Despite the fact that they are exceptionally capable and gifted in their exchange, they are individuals who can commit errors, uncommon however they may be. This may damagingly affect the business. PCs with refined AI similarity, then again, can take over monetary jobs and execute occupations precisely and with exactness, along these lines turning out to be preferred guardians over their human partners. Henceforth open trust reserves are gradually acquainting AI-based machines with keep command over reserves including observing and dynamic jobs.

12.4.13 Accounting Automation Avenues and Investment Management

There is no question that AI whenever executed appropriately will significantly affect the general working of any business remembering an ascent for efficiency and asset the executives. As of now, bookkeepers are utilizing different programming instruments and business measures the executives' apparatuses to show up at better-educated choices. As the innovation driving AI improves, more roads will open up to bookkeepers to robotize capacities in their calling that will additionally enhance business measures [29]. "Intelligent" speculation chiefs and mechanized abundance administrators can offer exact and precise monetary guidance, wiping out the requirement for full-time counsels and monetary experts. This has been a wellspring of much discussion among the worldwide venture local area. Truth be told, numerous enormous worldwide flexible investments have effectively decided on AI-based exchanging calculations that have totally removed the human component from market gauges and can foresee patterns dependent on recorded and measurable information. Be that as it may, if each financial backer were to utilize AI frameworks, it will be to the weakness of the whole market as it will essentially influence incomes and policymaking.

12.5 Challenges of AI Application in Accounting and Finance

"There is a way of thinking that predicts that all probably will not be well in the future in executing AI-based advances, and the psychological capacities that are looked to be bridled for better bookkeeping and different cycles may eventually have the option to produce self-ruling information and information" [27]. "As of now, the circumstance where AI can be utilized to control and alleviate adverse consequences is very far away. In the 2019 'EY Global FAAS' corporate detailing overview, 60% of Singapore respondents said the nature of money information delivered by AI cannot be trusted as much as information from regular money frameworks" [26]. The top dangers referred to comparable to transforming nonfinancial information into detailing data are keeping up with information protection, information security, and the absence of hearty information the board frameworks. Computer-based intelligence depends on admittance to immense volumes of information to be powerful, critical endeavors are subsequently expected to remove, change and house the information suitably and safely. The upside of AI frameworks is their capacity to break down and autonomously gain from different information and create important experiences. Nonetheless, this can be a two sided deal where an absence of appropriate information the executives or Cybersecurity frameworks can incline associations to huge dangers of incorrect experiences, information breaks, and digital assaults. Further, more modest associations might confront the issue of deficient information to construct models encompassing explicit regions for examination. Getting such information will likewise require frameworks and cycles to be set up and incorporated to guarantee that outer information outfit will supplement existing information. This requires critical monetary and time speculations. Thus, most organizations that carry out AI applications in their bookkeeping frameworks will probably zero in on regions that will have the hugest monetary and business impacts. This can be trying as more refined AI advancements are as yet in the outset stage and the main executions will consequently be probably not going to receive quick rewards. Indeed, even with the right information, there could in any case be a danger of AI calculation predisposition. On the off chance that the examples reflect existing predisposition, the calculations are probably going to intensify that inclination and may deliver results that build up existing examples of separation.

Another significant concern is the possible overexposure to digital related danger, programmers "who need to take individual information or classified data about an organization are progressively prone to target AI frameworks," given that these are not as adult or secure as other existing frameworks. While the enactment overseeing AI is as yet viewed as in their early stage that is set to change, frameworks that examine enormous volumes of purchaser information may not follow existing and unavoidable information protection guidelines and in this way, present dangers to associations. Likewise with any change drive, the human factor is basic to guaranteeing its prosperity. The advancement in AI advances is changing the jobs and obligations of bookkeepers, requiring capabilities past conventional specialized bookkeeping that additionally incorporate information on business and bookkeeping measures, including the frameworks supporting them. These capabilities are critical to adequately distinguish and apply use cases for AI advances, and work with compelling coordinated effort with different partners, including IT, lawful, assessment, and activities, during execution. In spite of these difficulties, the advantages of AI innovations stay convincing. The serious financial climate and fast innovative advances will drive reception. Over the long run, slow adopters will be disturbed and hazard becoming outdated. With the capability of AI innovations to be a distinct advantage for bookkeeping and money, reception is unavoidable and a sound AI methodology is vital to effective reception. While outfitting problematic innovations brings extraordinary freedoms, overseeing new dangers that accompany them is similarly as significant. Albeit the dangers rely upon each money capacity and individual application, associations should start by evaluating their circumstance against a range of potential dangers.

12.5.1 Data Quality and Management

This is the way to changing volumes of information into an association's essential resources. Associations ought to focus on building trust proactively in each aspect of the AI framework from the beginning. Such trust ought to stretch out to the essential reason for the framework, the honesty of information assortment and the executives, the administration of model preparing, and the thoroughness of strategies used to screen framework and algorithmic execution.

12.5.2 Cyber and Data Privacy

Contemplations ought to be made when planning and inserting AI advancements into frameworks. Creating legitimate framework partition and seeing how the framework handles the a lot of touchy information and settles on basic choices about people in a scope of regions, including credit, instruction, work, and medical care are basic to dealing with this danger.

12.5.3 Legal Risks, Liability, and Culture Transformation

At the most central level, associations need a careful comprehension of AI thinking and choices. There ought to likewise be components to permit an unmistakable review trail of AI choices and broad testing of the frameworks before sending. Hazard relief ought to likewise incorporate surveying the satisfactory expenses of mistake. Where the expenses of blunder are high, a human chief may in any case be expected to approve the yield to deal with this danger. As the innovation develops further, the worthy danger level can be changed as needs be. Fostering a fruitful AI execution guide requires recognizable proof and prioritization of utilization cases, with the arrangement that the human component is a principal piece of the condition. This is on the grounds that the interestingly human delicate abilities, like inventiveness and administration, just as human suspicion and judgment, are expected to address the new dangers that accompany the reception of arising advancements.

12.5.4 Practical Challenges

"Data volumes and quality are fundamental for the achievement of AI structures, without enough incredible data, models can basically not learn, restrictive accounting data is a lot of coordinated and unrivalled grade, and subsequently should be a promising early phase for making models" [29]. More unobtrusive affiliations probably will not have adequate data to enable accurate results, and basically, there may not be adequate data about undeniable issues to help extraordinary models. Mind blowing models may require external wellsprings of data, which may not by and large be practical to access at a fitting cost. Most importantly, AI is logically becoming consolidated into business and accounting programming. Therefore, various accountants will encounter AI without recognizing it, similar to how we use these capacities in our online looking or shopping works out. "This is the means by which more humble affiliations explicitly are likely going to take on AI instruments, second, perceptive gathering of AI abilities to handle unequivocal business or accounting issues will consistently require critical endeavor." While there is a huge load of free and open-source programming around here, the use of set up programming suppliers may be required for legitimate or authoritative reasons. Given the data volumes included, liberal gear and taking care of power may be required, whether or not it is gotten to on a cloud premise. In this manner, AI adventures will most likely focus in on districts that will have the best money related impact, especially cost decline openings, or those that are basic for significant arranging or customer support. "Various districts, while possibly profitable, may miss the mark on a strong theory case," also, using AI to encourage more vigilant things in master accounting districts may do not have the market potential to legitimize adventures from programming architects.

12.5.5 Limits of Machine Learning and AI

"While AI and ML models can be very mind blowing, there are as yet removed focuses to their abilities, and AI is surely not a general AI and models are not particularly versatile" [26]. Models sort out some way to do very certain tasks subject to a given course of action of data. Data sum and quality are fundamental, and not all issues have the right data to engage the machine to learn and many models require critical proportions of data. The tremendous forward jumps in areas like PC vision and talk affirmation rely upon outstandingly gigantic getting ready enlightening lists, extraordinary numerous data centers. "Yet that is not the circumstance with all spaces of AI, accomplishment depends after having satisfactory data of the right quality, and data regularly reflects existing inclination and predisposition in the public eye." Consequently, while models may conceivably forgo human inclinations, they can moreover get comfortable social tendencies that at this point exist. Also, a couple out of each odd issue will be sensible for an AI approach. For instance, there should be a level of repeatability about the issue so the model can sum up its learning and apply it to different cases. For special or novel inquiries, the yield might be undeniably less helpful. The yields of AI models are expectations or ideas dependent on numerical estimations, and not everything issues can be settled thusly. Maybe different contemplations ought to be calculated into choices, like moral inquiries, or the issue might require further underlying driver examination. Various degrees of prescient precision will likewise be proper in various conditions. It does not especially matter if proposal motors, for instance, produce wrong suggestions. Conversely, high levels of certainty are needed with clinical analysis or consistence undertakings. Giving express certainty levels close by the yield of models can be valuable choice guides in them. However, they accentuate the restrictions of models, the risks of improper dependence on them and the need to hold the contribution of people in numerous choice cycles.

12.5.6 Roles and Skills

"Affiliations will moreover expect induction to the right capacities, clearly, these beginnings with particular dominance in AI, yet, correspondingly similarly as with data examination, these specific capacities ought to be enhanced by significant appreciation of the business setting that incorporates the data and the agreement required" [25]. Accounting occupations are currently changing considering new capacities in data examination. Without a doubt, clerks are throughout set to work feasibly with data examination, as they combine critical levels of numeracy with strong business care. These examples will accelerate with AI. "A couple of occupations will continue to complement particular accounting capacity and human judgment to oversee problematic and novel cases, and various positions may develop to assemble participation and teaming up with various bits of the relationship to help them with getting the right importance from data and models." There will moreover be new situations, for example, accountants ought to be locked in with planning or testing models, or assessing estimations. They may need to take part in exercises to help with laying out the issues and fuse results into business measures. "Various clerks may be even more directly connected with managing the wellsprings of data or yields, for instance, exclusion dealing with or preparing data and this improvement will be reflected in the capacities expected of accountants." In any case capacities, accountants may need to adopt on better strategies for instinct and acting to benefit from AI devices.

12.5.7 Institutional Issues

"Bookkeeping has a more extensive institutional setting, and controllers and standard setters additionally need to construct their comprehension of the use of AI and be alright with any related dangers." Without this institutional help, it is beyond the realm of imagination to expect to accomplish change in regions like review or monetary announcing, subsequently, the dynamic contribution of standard setters and controllers here is fundamental. For instance, standard setters in review will need to analyze where examiners are utilizing these methods to acquire proof, and see how dependable the strategies are. Such bodies are now discussing the effect of information examination capacities on review norms, and thought of AI should expand on those conversations. "There are specific issues in this setting concerning the straightforwardness of models, in the event that associations and review firms progressively depend on discovery models in their tasks, seriously thinking will be needed with regards to how we acquire solace in their right activity." Controllers can likewise effectively empower and even push reception where it is adjusted to their work. "A significant part of the interest around here, for instance, is coming from monetary administrations associations to help administrative consistence and pressing factor from controllers."

12.6 Suggestions and Recommendation

To conquer protection from change and drive economical culture change, associations ought to infuse novel thoughts and new impulse into the group; one way is to recognize "change envoys" that are enabled by the executives to leave on new innovation drives and effective evidences of idea that would then be authorized for carry out to the association. Such comparative endeavors will be basic to defeat dormancy and obstruction and changing the money and bookkeeping ability blend might give a significant switch to culture change. By changing enrolment measures to support receptiveness and development, finance pioneers can try to draw in individuals from various areas and foundations who accompany new points of view and without the imbued suppositions and inclinations of run of the mill bookkeeping ability. Upskilling the current bookkeeping labor force past conventional money and bookkeeping abilities and reclassifying the profile for ability obtaining are key contemplations in driving a powerful computerized empowered labor force. The advantages of embracing AI advancements are obvious. While it is difficult to anticipate what AI advancements will at last mean for the bookkeeping business and calling, one thing is clear: organizations and bookkeeping experts need to contribute time in the near future to comprehend AI advances and environments, set out on verifications of idea to approve use cases, and drive social changes that adequately construct a genuinely computerized labor force and association for serious development.

Bookkeeping firms and bookkeepers ought to endeavor to work on their insight about AI as this will assist with upgrading their exhibition of different bookkeeping capacities, subsequently killing undesirable certain bookkeeping costs. There is potential for additional improvement through the utilization and advancement of more complicated AI applications, like neural organizations, master frameworks, fluffy frameworks, hereditary programming, and mixture frameworks and this chance ought to be researched to the furthest reaches conceivable. Digital guard ought to be fortified in other to satisfactorily ensure and uphold the framework's security and wellbeing. The executives ought to be accused of the obligation of guaranteeing that elective advances and specialists are reserve to offer specialized help benefits in the event of any breakdown or even to supplant any innovation that is broke down.

12.7 Conclusion and Future Scope of the Study

"The destiny of AI can should be a circumstance where machines will ultimately match individuals on various insightful planes, in any case today, it has made unprecedented types of progress and has viably shed occupations in the legal, banking and various endeavors" [24]. Accounting clearly has reliably absorbed new advancements and found ways to deal with get benefits from them. "Man-made intelligence should be no exclusion; it will not put accountants jobless yet will help them with inducing more business worth and capability from it" [28] between creating purchaser premium for modernized commitments, and the risk of taught new organizations, FIs are rapidly accepting progressed organizations by 2021; overall banks' IT monetary plans will flood to $297 billion. With ongoing school graduates and Gen Zers quickly transforming into banks' greatest addressable customer bundle in the US, FIs are being pushed to grow their IT and AI spending intends to satisfy higher modernized rules. The more youthful buyers incline toward advanced financial channels, with an enormous 78% of 20- to 30-year-olds never going to a branch if there is anything they can do about it. "And keeping in mind that the relocation from conventional financial channels to on the web and versatile banking was in progress pre-pandemic because of the developing chance among carefully local shoppers, the coronavirus drastically enhanced the move as stay-at-home requests were carried out the nation over and purchasers looked for more self-administration choices." "Insider Intelligence gauges both on the web and versatile financial reception among US buyers will ascend by 2024, coming to 72.8% and 58.1%, separately making AI execution basic for FIs appearing to be fruitful and serious in the advancing business."

References

1. Frey, C.B. and Osborne, M.A., The future of employment: How susceptible are jobs to computerization? *Technol. Forecast Soc Change*, *114*, 254–280, 2017.
2. Geissbauer, R., Vedso, J., Schrauf, S., Global Industry 4.0 Survey, in: *Industry 4.0: Building the digital enterprise*, pp. 5–6, 2016.
3. Piccarozzi, M., Aquilani, B., Gatti, C., Industry 4.0 in management studies: A systematic literature review. *Sustainability*, *10*, 10, 1–24, 20183821.
4. Milian, E.Z., Spinola, M.D.M., de Carvalho, M.M., Fintechs: A literature review and research agenda. *Electron. Commer. Res. Appl.*, *34*, 100833, 2019.

5. Arundel, A., Bloch, C., Ferguson, B., Advancing innovation in the public sector: Aligning innovation measurement with policy goals. *Res. Policy*, *48*, 3, 789–798, 2019.
6. Rikhardsson, P. and Yigitbasioglu, O., Business intelligence & analytics in management accounting research: Status and future focus. *Int. J. Account. Inf.*, *29*, 37–58, 2018.
7. Syrtseva, S., Burlan, S., Katkova, N., Cheban, Y., Pisochenko, T., Kostyrko, A., Digital Technologies in the Organization of Accounting and Control of Calculations for Tax Liabilities of Budgetary Institutions. *Stud. Appl. Econ.*, *39*, 7, 1–19, 2021.
8. Khan, A.K. and Faisal, S.M., The impact on the employees through the use of AI tools in accountancy. *Materials Today: Proceedings*, 2021.
9. Chandi, N., Accounting trends of tomorrow: What you need to know, 2018. https://www.forbes.com/sites/forbestechcouncil/2018/09/13/accounting-trends-of-tomorrow-what-you-need-to-know/?sh=744519283b4c [Date: 21/05/2022]
10. Ionescu, B., Ionescu, I., Tudoran, L., Bendovschi, A., Traditional accounting vs. Cloud accounting, in: *Proceedings of the 8th International Conference Accounting and Management Information Systems*, AMIS, pp. 106–125, 2013, June.
11. Christauskas, C. and Miseviciene, R., Cloud–computing based accounting for small to medium sized business. *Eng. Econ.*, *23*, 1, 14–21, 2012.
12. Schemmel, J., Artificial intelligence and the financial markets: Business as Usual?, in: *Regulating artificial intelligence*, pp. 255–276, Springer, Cham, 2020.
13. Syrtseva, S., Burlan, S., Katkova, N., Cheban, Y., Pisochenko, T., Kostyrko, A., Digital Technologies in the Organization of Accounting and Control of Calculations for Tax Liabilities of Budgetary Institutions. *Stud. Appl. Econ.*, *39*, 7, 1–19, 2021.
14. Yoon, S., A study on the transformation of accounting based on new technologies: Evidence from korea. *Sustainability*, *12*, 20, 8669, 2020.
15. Bauguess, S.W., The role of big data, machine learning, and AI in assessing risks: A regulatory perspective, in: *Machine Learning, and AI in Assessing Risks: A Regulatory Perspective*, SEC Keynote, OpRisk North America, 2017 June 21, 2017.
16. Cho, J.S., Ahn, S., Jung, W., The impact of artificial intelligence on the audit market. *Korean Acc. J.*, *27*, 3, 289–330, 2018.
17. Warren Jr., J.D., Moffitt, K.C., Byrnes, P., How big data will change accounting. *Account. Horiz.*, *29*, 2, 397–407, 2015.
18. IAASB, D., Exploring the Growing Use of Technology in the Audit, with a focus on data analytics, in: *Exploring the Growing Use of Technology in the Audit, with a Focus on Data Analytics*, 2016.

19. Bots, C.F.B., The difference between R=robotic process automation and artificialintelligence, 2018 May, 10, 2019. https://cfb-bots.medium.com/the-difference-between-robotic-process-automation-and-artificial-intelligence-4a71b4834788 [22/5/2022]
20. Davenport, T., Innovation in audit takes the analytics. AI routes, in: *Audit analytics, cognitive technologies, to set accountants free from grunt work*, 2016.
21. Chukwudi, O.L., Echefu, S.C., Boniface, U.U., Victoria, C.N., Effect of artificial intelligence on the performance of accounting operations among accounting firms in South East Nigeria. *Asian J. Economics, Bus. Account.*, 7, 2, 1–11, 2018.
22. Jędrzejka, D., Robotic process automation and its impact on accounting. *Zeszyty Teoretyczne Rachunkowośći*, 105, 137–166, 2019.
23. Ballestar, M.T., Díaz-Chao, Á., Sainz, J., Torrent-Sellens, J., Knowledge, robots and productivity in SMEs: Explaining the second digital wave. *J. Bus. Res., 108*, 119–131, 2020.
24. Greenman, C., Exploring the impact of artificial intelligence on the accounting profession. *J. Res. Bus. Econ. Manage., 8*, 3, 1451, 2017.
25. Kumar, K. and Thakur, G.S.M., Advanced applications of neural networks and artificial intelligence: A review. *Int. J. Inf. Technol. Comput. Sci., 4*, 6, 57, 2012.
26. Beerbaum, D., Artificial Intelligence Ethics Taxonomy-Robotic Process Automation (RPA) as business case. *Artificial Intelligence Ethics Taxonomy-Robotic Process Automation (RPA) as Business Case (April 26, 2021). Special Issue 'Artificial Intelligence& Ethics' European Scientific Journal*, 2021.
27. Shubhendu, S. and Vijay, J., Applicability of artificial intelligence in different fields of life. *Int. J. Sci. Eng. Res., 1*, 1, 28–35, 2013.
28. Taghizadeh, A., Mohammad, R., Dariush, S., Jafar, M., Artificial intelligence, its abilities and challenges. *Int. J. Bus. Behav. Sci., 3*, 12, 2013.
29. Gusai, O.P., Robot human interaction: Role of artificial intelligence in accounting and auditing. *Indian J. Account, 51*, 1, 59–62, 2019.

13

Obstacle Avoidance Simulation and Real-Time Lane Detection for AI-Based Self-Driving Car

B. Eshwar*, Harshaditya Sheoran, Shivansh Pathak and Meena Rao

Department of ECE, Maharaja Surajmal Institute of Technology, Janakpuri, New Delhi, India

Abstract

This chapter aims at developing an efficient car module that makes the car drive autonomously from one point to another avoiding objects in its pathway through use of Artificial Intelligence. Further, the authors make use of visual cues to detect lanes and prevents vehicle from driving off road/moving into other lanes. The paper is a combination of two simulations; first, the self-driving car simulation and second, real-time lane detection. In this work, Kivy package present in Anaconda navigator is used for simulations. Hough transformation method is used for lane detection in "restricted search area."

***Keywords*:** Self-driving car, artificial intelligence, real-time lane detection, obstacle avoidance

13.1 Introduction

A self-driving car is designed to move on its own with no or minimal human intervention. It is also called autonomous or driverless car many times in literature [1]. The automotive industry is rapidly evolving and with it the concept of self-driving cars is also evolving very fast. Several companies are focused on developing their own self. Even the tech giants, which are not into "mainstream automobile," like Google and Uber, seem

**Corresponding author*: b.eshwar13@gmail.com

M. Niranjanamurthy, Kavita Sheoran, Geetika Dhand, and Prabhjot Kaur (eds.) Data Wrangling: Concepts, Applications and Tools, (275–288) © 2023 Scrivener Publishing LLC

greatly interested in it. This is due to the ease of driving opportunity that self-driving cars provide. The self-driven cars that make use of artificial intelligence to detect the obstacles around and run in an auto pilot mode are a major area of research and study these days [2]. The self-driving cars allow the users to reach their destination in a hassle free manner giving complete freedom to undertake any other task during the time of travel. Moreover, human involvement is also least, and hence, the chance of human error leading to accidents also minimizes in self-driving cars. In driverless cars, people sitting in the car would be free of any stress involved in driving and on road traffic. However, to make the self-driving cars a common phenomenon, various features have to be developed, and the system should be developed in such a way that the driverless car is able to navigate smoothly in the traffic, follow lanes and avoid obstacles. Researchers have worked across different techniques and technologies to develop the system [3]. An autonomous platform for cars, using the softmax function, is presented, which gives out the outputs of each unit between 0 and 1. The system only uses a single camera [4]. Further research was carried out in real-time in order to find the positions on the roadway by Miao *et al.* Canny edge extraction was administered so as to obtain a map for the matching technique and then to select possible edge points [5]. In literature, an autonomous RC car was also proposed and built making use of artificial neural network (ANN). Fayjie *et al.* in their work have implemented autonomous driving using the technique of reinforcement-learning based approach. Here, the sensors used are "lidar" that detects objects from a long distance [6]. The simulator used gimmicks real-life roads/traffic. Shah *et al.* used deep neural to detect objects. Prior to their work, "conventional deep convolution neural network" was used for object detection. Yoo *et al.* had proposed a method that creates a new gray image from a colored image formulated on linear discriminant analysis [7]. Hillel *et al.* elaborated and tried to tackle various problems that are generally faced while detecting lane like image clarity, poor visibility, lane and road appearance diversity [8]. They made use of LIDAR, GPS, RADAR, and other modalities to provide data to their model. Further using obstacle detection, road and lane detection was done and details were fed to the vehicle to follow in real-time. In the work by Gopalan *et al.*, the authors discuss the most popular and common method to detect the boundaries of roads and lanes using vision system [9]. General method to find out different types of obstacles on the road is inverse perspective mapping (IPM). It proposes simple experiment that is extremely effective in both lane detection and object detection and tracking in video [10]. Clustering technique has also been used to group the detected points [11]. Results were found to be effective in terms of detection and tracking of multiple vehicles at one time irrespective of the distance involved.

The authors of this chapter were motivated by the work done by earlier researchers in the domain of self-driving. The objective of this work presented in the chapter is to develop a model of a car that detects lane and also avoid obstacles. Lane detection is a crucial component of self-driving cars. It is one among the foremost and critical research area for understanding the concept of self-driving. Using lane detection techniques, lane positions can be obtained. Moreover, the vehicle will be directed to automatically go into low-risk zones. Crucially, the risk of running into other lanes will be less and probability of getting off the road will also decrease. The purpose of the proposed work is to create a self-driving car model that could sustain in traffic and also avoids accidents.

13.1.1 Environment Overview

13.1.1.1 Simulation Overview

The self-driving car application uses Kivy packages provided in anaconda navigator. The aim is to allow for speedy as well as easy interactive design along with rapid prototyping. Also, the code should be reusable and implementable. The application environment in which the car "insect" will appear is made using the Kivy packages. The environment will have the coordinates from 0,0 at the top left to 20,20 at bottom right and the car "insect" will be made to traverse from the bottom right to the top left i.e. these will be the source and destination.

The idea/motive of creating this is that the car learns not only to traverse from source to destination but at the same time avoids the obstacles. These obstacles should be such that the user/developer should be able to draw and redraw the pathway for the agent as and when the agent learns the given pathway to destination and back to source. Also, the pathways, thus, created must also provide the punishment to the agent if the agent hits it. With rewards and punishments is how the agent learns. Alongside this very approach to provide punishment of to the agent as it touched the pathway or the obstacle, the degree of the punishment should vary depending upon the thickness of the pathway. Hence, the tracks that could be drawn was made such that holding the mouse pointer for longer period of time increased the thickness. The greater the thickness, the more the punishment.

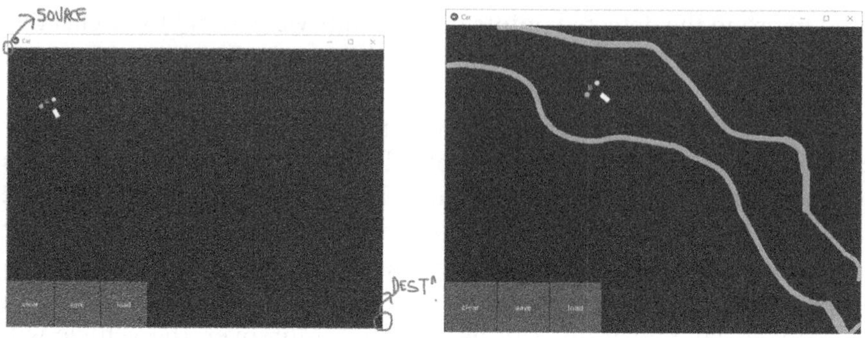

Figure 13.1 Self-driving car UI.

Since the simulation requires a need to draw and redraw the pathways as and when the agent learns the path, there is a "clear" button that clears the tracks that were created till then refer Figure 13.1.

13.1.1.2 Agent Overview

The agent created is designed to have three sensors. The sensors are placed one right at the front center, the rest two at 20 degrees to the left and right of the center sensor, respectively. These sensors can sense any obstacle which falls under the + −10 degree sector from the center of axis of the particular sensor. This is an added rectangular body just for representation, it has no functionality as such. The rectangular body gives the coordinate where the agent exists. The body moves forward, moves right left at 10-degree angle.

The sensors when finds no obstacle in front of them, it updates the information and moves in a random direction as it is exploring. Depending on the reward or the punishment that it receives upon the action that it took, it learns and takes a new action. Once the car "agent" reaches the goal it earns a reward of +2. Punishment value is decreased for going further away from the goal since avoiding the sand sometimes requires agent to move away from the destination.

Cumulative reward is introduced, instead of giving it a certain value, independent conditions sum up their rewards (which mostly are penalties). Hitting the sand earns the agent a negative reward of –3. Penalty for turning is also introduced. The model should keep its direction in more conservative way. Replacement of integral rewarding system with a binary reward for closing to the target with continuous differential value. This lets the brain keep direction, this reward is really low, yet still a clue for taking proper action for the brain.

13.1.1.3 Brain Overview

This application also uses NumPy and Pytorch packages for deep learning and establishment of neural networks that define what actions are to be taken depending upon the probability distribution of reward or punishment received. Numpy is a library that supports massive, multidimensional arrays and matrices, as well as a large number of complicated mathematical functions to manipulate them. PyTorch contains machine learning library, which is open source, and it is used for multiple applications.

13.1.2 Algorithm Used

The agent designed uses Markov decision process and implements deep q-learning and a neural network along with a living penalty added so that the agent does not only keep moving in the same position but also reaches the destination.

13.1.2.1 Markovs Decision Process (MDP)

Markovs decision process (MDP), derived from bellman equation, offers solutions with finite state and action spaces. This is done by techniques like dynamic programming [12]. To calculate optimal policies value, which contains real values, and policy, which contains actions, are stored in two arrays indexed by state. We will acquire the solution as well as the discounted sum of the rewards that will be won (on average) by pursuing that solution from state at the end of the algorithm. The entire process can be explained as updating the value and a policy update, which are repeated in some order for all the states until no further changes happen [13]. Both recursively update a replacement of the optimal policy and state value using an older estimation of these values.

$$V(s)=\max_a\left(R(s,a)+\gamma\sum_{s'}P(s,a,s')V(s')\right) \tag{13.1}$$

V(s) is the value or reward received by the agent for taking an action "a" in state "s."

Here, the order is based on the type of the algorithm. It can be done for all states at one time or one by one for each state. To arrive at the correct solution, it is to be ensured that no state is permanently excluded from either of the steps.

13.1.2.2 Adding a Living Penalty

The agent resorts to keep bumping at the corner walls near the state where there is −1 reward. It learns that by keep bumping on the wall it will not receive a punishment but due to it being a MDP it does not know yet that +2 reward is waiting if it makes to the destination. Living penalty is a punishment or negative reward given to the agent and after appropriate simulations so that the penalty is not high enough to force the agent to directly fall into the wall since the reward is too low for it to keep trying to find the right action. At the same time, punishment should not be small enough to let the agent remain in same position. Q-learning is a "model-free reinforcement learning algorithm." It basically defines or suggest what action to be taken under different situations [14].

$$Q(s,a) = R(s,a) + \gamma \sum_{s'} (P(s,a,s')V(s')) \tag{13.2}$$

$Q\ (s,\ a)$ is the quality of taking the action a at the state s after calculating the cumulative value or reward on being on next state s'. This is derived from MDP.

13.1.2.3 Implementing a Neural Network

When designing the agent environment to the agent is described in terms of coordinates, i.e., vectors and supplying the coordinates to the neural network to get appropriate Q values [15]. The neural network (NN) will return 4 Q values (up, down, left, right). These will be the target Q values that the model has predicted before agent performs any action and are stored.

$$TD(a,s) = R(s,a) + \gamma \max_{a'} Q(s',a') - Q(s,a) \tag{13.3}$$

$$Q\ (s,\ a) = Q\ (s,\ a) + \alpha TD\ (a,\ s) \tag{13.4}$$

Now when the agent actually performs the actions randomly and gets Q value, it is compared with the targeted values and the difference is called temporal difference (TD). TD is generally intended to be 0 or close to 0 i.e. the agent is doing what is predicted or learnt. Hence, the loss is fed as the input to the NN to improve the learning.

13.2 Simulations and Results

13.2.1 Self-Driving Car Simulation

Creation of more challenging path design was done that will better train the agent to traverse difficult paths and still reach the destination. The more maze-like pathways better the agent learns, hence research on the pathways that generally are used to train such models was explored. The designs that were worked upon had also to be accurate in real world otherwise it becomes more like a game than real world application worthy. Improvization in independent cars are ongoing and the software within the car is continuously being updated. Though the development started with the module of driver free cars, it has now progressed to utilizing radiofrequency, cameras, sensors, more semiautonomous feature and in turn reducing the congestion, increasing safety with faster reactions, and fewer errors. Despite all of its obvious advantages, autonomous car technology must also overcome a slew of social hurdles. Authors have tried to simulate the self-driven car for various difficult tracks and different situations Figure 13.2 shows a simple maze track with no loops involved. On the other hand, Figure 13.3 shows simulation in a hair pin bend and shows a more difficult path with multiple loops. Figure 13.4 for more difficult path to cope with looping paths.

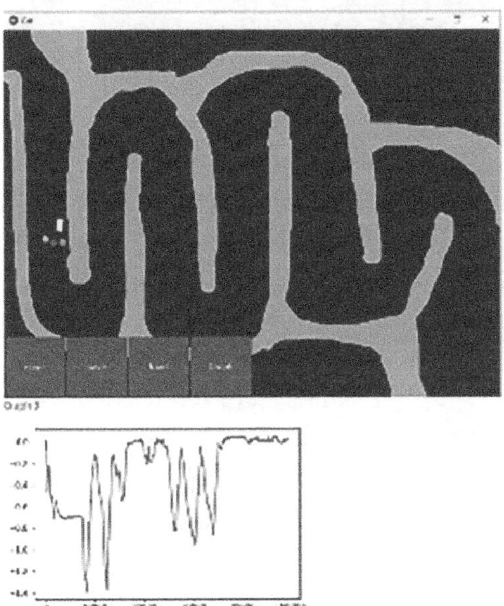

Figure 13.2 Simple Maze with no to-fro loops involved.

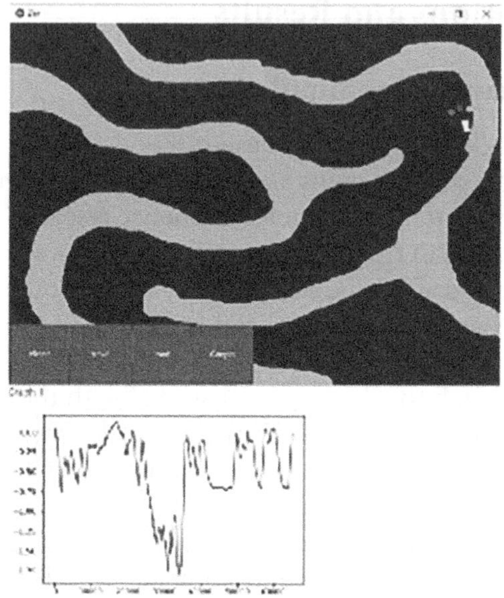

Figure 13.3 Teaching hair-pin bends.

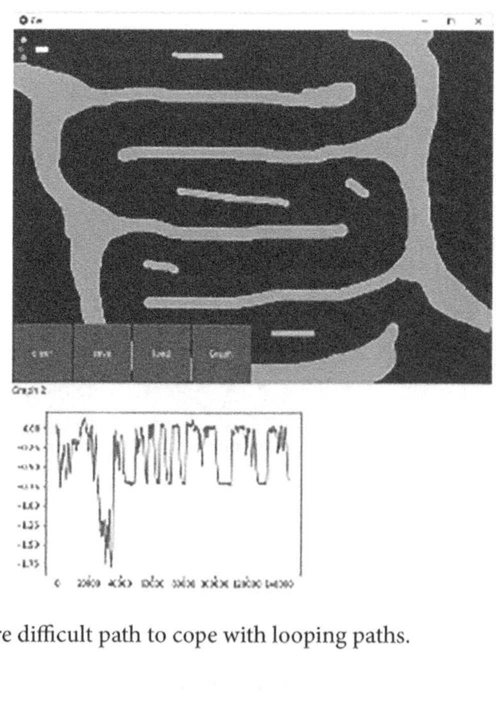

Figure 13.4 A more difficult path to cope with looping paths.

13.2.2 Real-Time Lane Detection and Obstacle Avoidance

A lane is designated to be used by a series of vehicles, to regulate and guide drivers and minimize traffic conflicts. The lane detection technique uses the OpenCV, image thresholding and Hough transform. It is a solid line or a much rugged/dotted line that identifies the positioning relationship between the lane and the car. Lane detection is a critical aspect of the driver free cars. An enhanced Hough transform is used to enable straight-track lane detection, whereas the tracking technique is investigated for curved section detection [16]. Lane detection module makes use of the frames that are provided to it by breaking any video of a terrain/road taken into frames and detects lanes. This entire process is explained through the flow-chart shown in Figure 13.5. The lanes are detected, and marking is made on those frames. These frames are then stitched together again to make an mp4 video output, which is the desired result.

13.2.3 About the Model

This module makes use of OpenCV [17]. The library is used mainly for "image processing," "capturing video" and its analysis including structures like face/object detection. Figure 13.6 depicts a lane and Figure 13.7 depicts the lane detection from video clips.

Each number represents the pixel intensity at a specific site. Figure 13.8 provides the pixel values for a grayscale image with a single value for the intensity of the black color at that point in each pixel. The color images will

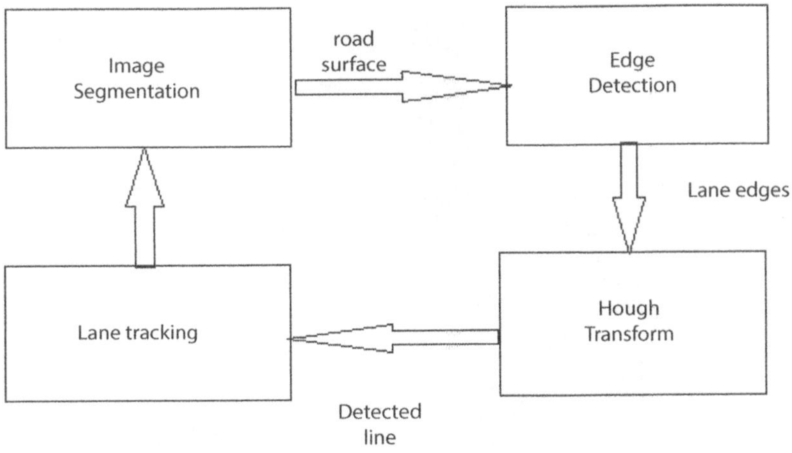

Figure 13.5 Plan of attack to achieve the desired goal.

Figure 13.6 Lane.

Figure 13.7 Lane detection from video clips.

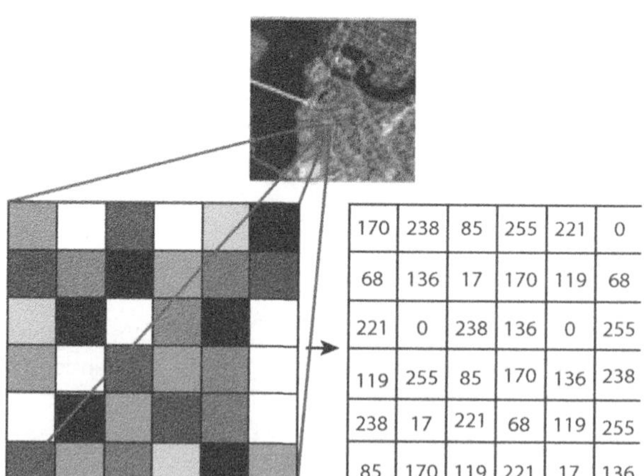

Figure 13.8 Depiction of pixel value.

Figure 13.9 Setting area of interest on the frame.

have various values for one pixel. These values characterize the intensity of respective channels—red, green, and blue channels for RGB images.

In a general video of a car traversing on a road, there are various things in any scenario apart from the traditional lane markings. There are automobiles on the road, road-side barriers, street-lights, etc. In a video, scenes changes at every frame and this reflects actual driving situations pretty well. Prior to resolving the lane detection issue, ignoring/removal of the unwanted objects from the driving scene is done [18]. The authors have narrowed down the area of interest to lane detection. So, instead of working with the entire frame, only a part of the frame will be worked upon. In the image below, apart from the lane markings (already on the road), everything else like cars, people, boards, signals etc. has been hidden in the frame. As the vehicle moves, the lane markings would most likely fall in this area only. Figure 13.9 shows how area of interest is set on the frame.

13.2.4 Preprocessing the Image/Frame

First, the image/frame is processed by masking it. NumPy array acts as a frame mask. The technique of applying mask to an image is that, pixel values of the desired image is simply changed to 0 or 255 or any other number [19]. Second, thresholding is applied on the frame. Here, provides the pixel values for a grayscale image with a single value for the intensity of the black color at that point in each pixel. The pixel can be assigned any one of the two values depending on whether the value of the pixel is greater than or

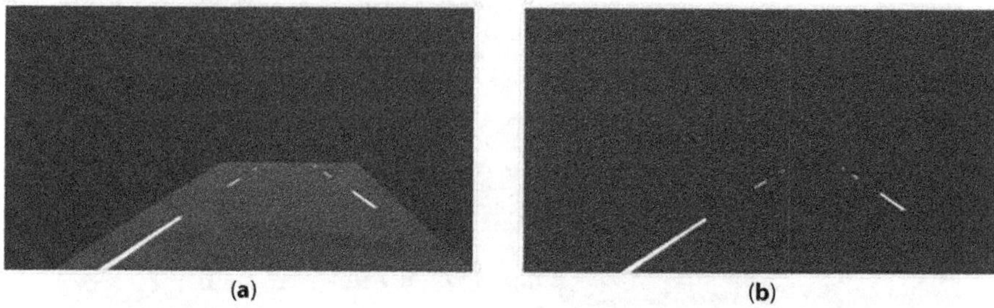

Figure 13.10 (a) Masked image (b) Image after thresholding.

lower than the threshold value. Figures 13.10 (a) and (b) show a masked image and image after thresholding respectively.

When threshold is applied on the masked image, there is only lane markings in the output image. Detecting these lane markings are done with the help of "Hough Line Transformation" [20]. In this work, the objective is to detect lane markings that can be represented as lines. Finally, the above process performed on a single frame from a video is repeated on each frame and each frame is then compiled in the form of a video. This gives the final output in a Mp4 video format.

13.3 Conclusion

The images in Figure 13.11 show the detection of lanes on various video frames. By detecting lanes, the self-driving car will follow a proper route and also avoid any obstacle.

Figure 13.11 Shows lane detection in various frames of the video.

In this work, a "real-time lane detection algorithm based on video sequence" taken from a vehicle driving on highway was proposed. The proposed model uses a series of images/frames snapped out of the video. Hough transformation was used for detection of lanes with restricted search area. The authors were also able to demonstrate the simulation of a self-driving car on easy as well as difficult mazes and tracks. Subsequently, the lanes were also detected on various frames. The lane detection helps the self-driving car to move on the track while avoid any obstacles. In this way, self-driving through tack along with lane detection and obstacle avoidance was achieved.

References

1. By: IBM Cloud Education, What is Artificial Intelligence (AI)? *IBM*. Available: https://www.ibm.com/cloud/learn/what-is-artificial-intelligence.
2. de Ponteves, H., Eremenko, K., Team, S.D.S., Support, S.D.S., Anicin, L., Artificial Intelligence A-Z™: Learn how to build an AI. *Udemy*. Available: https://www.udemy.com/course/artificial-intelligence-az/.
3. Seif, G., Your guide to AI for self-driving cars in 2020. *Medium*, 19-Dec-2019. Available: https://towardsdatascience.com/your-guide-to-ai-for-self-driving-cars-in-2020-218289719619.
4. Omrane, H., Masmoudi, M.S., Masmoudi, M., Neural controller of autonomous driving mobile robot by an embedded camera. *2018 4th International Conference on Advanced Technologies for Signal and Image Processing (ATSIP)*, 2018, doi: 10.1109/atsip.2018.8364445.
5. Miao, X., Li, S., Shen, H., On-board lane detection system for intelligent vehicle based on monocular vision. *Int. J. Smart Sens. Intell. Syst.*, *5*, 4, 957–972, 2012, doi: 10.21307/ijssis-2017-517.
6. Shah, M. and Kapdi, R., Object detection using deep neural networks. *2017 International Conference on Intelligent Computing and Control Systems (ICICCS)*, 2017, doi: 10.1109/iccons.2017.8250570.
7. Yoo, H., Yang, U., Sohn, K., Gradient-enhancing conversion for illumination-robust lane detection. *IEEE Trans. Intell. Transport. Syst.*, *14*, 3, 1083–1094, 2013, doi: 10.1109/tits.2013.2252427.
8. Hillel, A.B., Lerner, R., Levi, D., Raz, G., Recent progress in road and lane detection: A survey. *Mach. Vis. Appl.*, *25*, 3, 727–745, 2012, doi: 10.1007/s00138-011-0404-2.
9. Gopalan, R., Hong, T., Shneier, M., Chellappa, R., A learning approach towards detection and tracking of lane markings. *IEEE Trans. Intell. Transport. Syst.*, *13*, 3, 1088–1098, 2012, doi: 10.1109/tits.2012.2184756.
10. Paula, M.B.D. and Jung, C.R., Real-time detection and Ccassification of road lane markings[C]. *Xxvi Conference on Graphics Patterns and Images*, pp. 83–90, 2013.

11. Kaur, G., Kumar, D., Kaur, G. *et al.*, Lane detection techniques: A Review[J]. *Int. J. Comput. Appl.*, 4–6, 112.
12. Stekolshchik, R., How does the Bellman equation work in Deep RL? *Medium*, 16-Feb-2020. Available: https://towardsdatascience.com/how-the-bellman-equation-works-in-deep-reinforcement-learning-5301fe41b25a.
13. Singh, A., Introduction to reinforcement learning: Markov-decision process. *Medium*, 23-Aug-2020. Available: https://towardsdatascience.com/introduction-to-reinforcement-learning-markov-decision-process-44c533ebf8da.
14. Violante, Simple reinforcement learning: Q-learning. *Medium*, 01-Jul-2019. Available: https://towardsdatascience.com/simple-reinforcement-learning-q-learning-fcddc4b6fe56.
15. Do, T., Duong, M., Dang, Q., Le, M., Real-time self-driving car navigation using deep neural network. *2018 4th International Conference on Green Technology and Sustainable Development (GTSD)*, 2018, doi: 10.1109/gtsd.2018.8595590.
16. Qiu, D., Weng, M., Yang, H., Yu, W., Liu, K., Research on lane line detection method based on improved hough transform. *Control And Decision Conference (CCDC) 2019 Chinese*, pp. 5686–5690, 2019.
17. About, OpenCV, in: *OpenCV*, 04-Nov-2020, Available: https://opencv.org/about/.
18. Guidolini, R. *et al.*, Removing movable objects from grid maps of self-driving cars using deep neural networks. *2019 International Joint Conference on Neural Networks (IJCNN)*, 2019, doi: 10.1109/ijcnn.2019.8851779.
19. Image Masking with OpenCV. *PyImageSearch*, 17-Apr-2021. Available: https://www.pyimagesearch.com/2021/01/19/image-masking-with-opencv/.
20. Hough Line Transform. *OpenCV*. Available: https://docs.opencv.org/3.4/d9/db0/tutorial_hough_lines.html. [12/11/2021].

14

Impact of Suppliers Network on SCM of Indian Auto Industry: A Case of Maruti Suzuki India Limited

Ruchika Pharswan[1]*, Ashish Negi[2] and Tridib Basak[3]

[1]Bharti School of Telecommunication Technology and Management, Indian Institute of Technology, Delhi, New Delhi, India
[2]Department of Electronics and Communication Engineering, HMR Institute of Technology and Management, Hamidpur, New Delhi, India
[3]Department of Computer Science Engineering, HMR Institute of Technology and Management, Hamidpur, New Delhi, India

Abstract

Maruti Suzuki India Limited (MSIL) has been the most fascinating story among automobile manufacturing enterprises, and it is India's largest car manufacturer. After Maruti merged with Suzuki, it acquired high acceleration in automaker industry. MSIL has a vast network of vendor deals and service networks across the country and primarily focuses on providing cost-effective products with high customer satisfaction. The proposed report is single case analysis and aims to provide a comprehensive view of goal, strategic perspectives, and various aspects implemented in the supply chain, inventory, logistics management, and the benefits inferred by MSIL in order to gain a competitive advantage. We have also tried to figure out how the current epidemic (COVID-19) has affected their SCM and how they have adapted their business strategy to deal with it. This case study reveals that MSIL has been working hard to improve its supply chain and logistics management in order to achieve positive results.

Keywords: Automotive industry, Maruti Suzuki India Limited, COVID-19, supply chain management

**Corresponding author*: Ruchi1996pharswan@gmail.com

M. Niranjanamurthy, Kavita Sheoran, Geetika Dhand, and Prabhjot Kaur (eds.) Data Wrangling: Concepts, Applications and Tools, (289–314) © 2023 Scrivener Publishing LLC

14.1 Introduction

Automotive industry is one of the predominant pillars of the growing Indian economy and, to a huge extent, serves as a bellwether for its modern state. In 2018 alone, the automotive sector contributed 7.5% of total gross domestic product (GDP) of India. Owing to COVID-19 this year, this percentage dipped to 7%. The Government of India and automotive industry experts expect it to transpire as the worldwide third-largest passenger vehicle market by the end of 2021 with an increase of 5% from existing percentage and draw US $ 8 to 10 billion local and foreign investment by 2023. In fiscal year 2016-2020 (FY16-20), annual growth rate of the Indian automotive market was over 2.36% compound annual growth rate (CAGR), indicating a positive trend in step forward [1]. The automotive industry is likely to generate five crore direct and indirect jobs by 2030. India Energy Storage Alliance (IESA), published its 2nd annual "India Electric Vehicle Market Overview Report 2020–2027" on the Indian Market, which states that in India during FY 19-20, EV sales stood at 3,80,000, and on the other hand the EV battery market accounted for 5.4 GWh, pointing to the growth of the Indian EV market at a CAGR of 44% between 2020 and 2027. In FY20, India was the fifth largest auto market worldwide [2]. And in 2019, India secured the seventh position among top 10 nations, in commercial vehicle manufacturing. Recent reports show that in April to June 21 Indian automotive exports stood at 1,419,439 units, which is approximately three times more than that of the export of 436,500 units during the same period last year [3].

Starting from the era, when Indian manufacturers disbursed upon foreign ties, to now developing their own innovation, the Indian auto sector has come a long way. While looking up for top 10 automobile players in India, Maruti Suzuki has always been on top of the table [4]. A consistent dominant leader standing deep rooted with big names like Tata Motors, Hyundai Motors, Toyota, Mahindra & Mahindra, and Kia. With a good start to 2021, Maruti Suzuki India Ltd topped the four wheelers chart with a 45.74% market cap. MSIL was formerly known as Maruti Udyog Limited, and is a subsidiary of Suzuki Motor corporation primarily known for its services, was founded by the Indian Government in 1981 [5]. Later on, it was sold to Suzuki Motor Corporation in 2003. Hyundai with a 17.11% market share ended in second place. Hyundai Motor India Ltd (HMIL) is a proprietary company of Hyundai Motor Company, established in 1998 and is headquartered in Chennai, Tamil Nadu. It deals across nine models across segments and exports to nearly 88 countries across the globe. Tata

Motors, the biggest Gainers of January 2021 bagged third position with a market share of 8.88%. While Mahindra & Mahindra sold 20,498 units against 19,55 units in 2020 with a market share of 6.27%.Whereas, Honda, Kia, Nissan, and Toyota bagged 3.72%, 6.27%, 1.49%, and 2.70% market cap, respectively.

As a matter of fact, over the past two decades, the global auto industry sales have declined almost 5%, that is approximately down to less than 92.2 million vehicles. These, however, are very different from the declines that the companies in the industry have seen since 2019 owing to COVID-19. In a report by Boston Consulting group, it highlights a wide range of actions, including revitalizing the supply chain, cost reduction in operations, and reinventing user-based makeovers in the marketing strategies that were adopted by various players of the auto sector, which made them survive the pandemic. Another article by KPMG draws one's attention to India-specific strategies, such as localization and sustainability of supply chain, mobilization of marketing strategies and growth of subscription models, such as virtual vehicle certification, which helped to revive the Indian automobile industry. Keeping a close eye on the strategies followed by the most of the stakeholders can help us to admit the choice pattern why these firms walked a specific strategy and bounced back post COVID-19 [7]. Thus, stakeholders may be able to seek a series of incremental strategies than those who can be a pause from the past. The International Organization of Motor Vehicle Manufacturers, a.k.a. "Organization Internationale des Constructeurs d'Automobiles" (OICA) in its report mentions that a 16% decline of 2020 global automobile production has pushed back up to 2010 equivalent sales levels. Europe, which represents an almost 22% share of global production, dipped more than 21%, on average ranging from 11% to almost 40% across the European countries. And Africa on the other hand has also faced a sharp decline of more than 35%. Meanwhile, America, which upholds 20% share of global production, dropped by 19%. Moving to the south, the South America continent declined by more than 30%, whereas Asia declined with only 10% even after the fact that it is the world's largest manufacturing region, with a market share of 57% global production [8].

While India's automotive sector has experienced numerous hurdles in recent years, including the disastrous COVID-19 pandemic, it continues to thrive, and has made its way through most of the challenges and many are now in the rear-view mirror [9]. From global supply-chain rebalancing, an outlay of ₹ 26,058 crore Government incentives boosting exports and high-value advance automotive technology and technology disruptions creating white spaces have created opportunities at all stages of local automotive value chain strategies. Globally, a few original equipment manufacturers (OEMs) have started showing their presence in downstream value chain ventures like BMW's secure assistance now offering finance and insurance services. Ford's agreement with GeoTab opened its door to the vehicle data value chain [10]. Even in India, experiments like iAlert, e-diagnostics, Service Mandi by Ashok Leyland, True valued by Maruti Suzuki in downstream ventures have provided opportunities to shape a digitally enabled ecosystem. Which provided a comprehensive solution creating a world-class ownership experience, with services like scheduled services, breakdown service, resale, or purchase. Innovative brands, like Tesla, expect the fact that going through digital channels is the future against traditional brick and mortar channels [11]. In view of MSIL's experiences in the Indian automotive business, this current study aims to investigate and broaden the horizon by examining the environment for factors that contributed to MSIL's long-term viability when other important participants were unable to, both during and after the COVID-19 pandemic. Various aspects implemented in the supply chain, inventory, and logistics management, benefited MSIL. And the strategic viewpoints that propelled MSIL to the forefront of the Indian automotive market [12].

The remaining sections of this study are organized as follows: Section 14.2 presents the multiple perspectives and researches on Automotive Industry from the expert contributors and overall themes within literature. Section 14.3 exhibits the workflow and methods used in this case study. Section 14.4 details the key findings and statistics of the study using secondary resources [5]. Section 14.5 depicts the discussion on the key automotive industry related topics relating to the challenges, opportunities and the research agenda presented by the expert contributors. The study is concluded in section 14.6.

14.2 Literature Review

The automotive sector is rapidly developing and integrating cutting-edge technologies into its spectrum. We reviewed a number of research papers and media house articles/publications and selected the ones that were related to our study and overall themes in the literature [13].

In their research paper, M. Krishnaveni and R. Vidya illustrated the growth of the Indian automobile industry. They looked into how the globalization process has influenced the sector in terms of manufacturing, sales, personal research and development, and finance in their report. They also came to the conclusion that, in order to overcome the challenges provided by globalization, Indian vehicle makers must ensure technological innovation, suitable marketing tactics, and an acceptable customer care feedback mechanism in their businesses [14]. The impact of COVID19 on six primary affected sectors, including automobiles, electricity and energy, electronics, travel, tourism and transportation, agriculture, and education, has been highlighted in the article, authored by Janmenjoy Nayak and his five fellow mates [15]. They also looked at the downstream effects of the automobile sector, such as auto dealers, auto suppliers, loan businesses, and sales, in their report. They also mentioned some of the difficulties that have arisen as a result of COVID-19, such as crisis management and response, personnel, operations and supply chain, and financing and liquidity [16].

Shuichi Ishida examines in her research, how product supply chains should be managed in the event of a pandemic using examples from three industries: automobiles, personal computers (PCs), and household goods. In their study, it was found that vehicle production bases had been transformed into "metanational" firms, whereas earlier they had built a primarily local SCN center on the company's home location [17]. As a result, in the future, switching to a centralized management style that takes advantage of the inherent strength of a "closed-integral" model, which maximizes the closeness of suppliers to manufacturing sites, would be advantageous. The study by Zhitao Xu and his fellow researchers intends to investigate the COVID-19 impacts on the efficacy and responsiveness of global supply chains and provide a set of managerial insights to limit their risks and strengthen their resilience in diverse industrial sectors using critical reading and causal analysis of facts and figures [18]. In which they stated that global output for the automotive sector is anticipated to decrease by 13%. Volkswagen halted its vehicle facilities in China due to travel restrictions and a scarcity of parts. General Motors restarted its Chinese facilities for the same reasons, although at a relatively modest manufacturing pace. Due to a shortage of parts from China, Hyundai's assembly plants in South Korea were shut down. Nissan's manufacturing sites in Asia, Africa, and the Middle East have all shut down [19].

In their study paper, Pratyush Bhatt and Sumeet Varghese described the current state of the automobile sector and how it may strategize in the face of economic uncertainty [20]. They pointed out that material expenses (which are the greatest in absolute terms compared to the rest) have risen from 56.3 percent to 52.3 percent to a quick increase of 62.6 percent, resulting in a relative increase of 0.4 percent over three years, thanks to steady investment in people. Avoiding the need for an intermediary between the company and the client, as well as preparing deliveries to arrive at the customer directly from the service centre, are two more cost-cutting measures

(Maruti Suzuki Readies Strategy, n.d.). As a result, in order to reverse the profit decline trend, they should contemplate proportionately divesting in both divisions while maintaining their borrowing pattern, which is "keeping it less." Manjot Kaur Shah and Sachin Tomer, in their research paper discussed how different businesses in India interacted with the public during COVID-19 in order to preserve a healthy relationship with their fan base as a marketing strategy [4]. Automobile manufacturers' brands were also emphasized in this study. Maruti Suzuki, for example, advised customers not to drive during the shutdown and to stay inside. #FlattenTheCurve, #GearUpForTomorrow, and #BreakTheChain were among the hashtags used. Furthermore, they made a contribution by distributing 2 million face masks. Hyundai was a frequent Instagram user. They urged their followers to be safe, emphasizing that staying at home is the key to staying safe [21]. #HyundaiCares, #WePledgeToBeSafe, and #0KMPH were among the hashtags they used. People were also instructed to take their foot off the pedal and respect the lockdown. The first post from Toyota India was made on March 21, 2020, ahead of a one-day shutdown in India on March 22, 2020. They used the hashtag #ToyotaWithIndia to show that Toyota is standing with India in its fight against COVID-19. Hero MotoCorp has extended their guarantee until June 30, 2020. They offered advice on how to keep bikes when they are not in use. #Stayhomestaysafe was one of the hashtags they used [22].

14.2.1 Prior Pandemic Automobile Industry/COVID-19 Thump on the Automobile Sector

COVID-19 was proclaimed as a global pandemic by the World Health Organization (WHO) as soon as it was found, a lot of industries have been affected by the same including the Automobile sector worldwide and in India as well. The worldwide epidemic caused by the coronavirus struck at a time when both the Indian economy and the Automobile industry were anticipating recuperation and firm growth [23]. While the GDP gain forecasts were expected to be scaled by 5.5%, the pandemic resulted in a negative impact of 1-2% on the awaited magnification rates for the same .In India, the introduction of Covid-19 had a negative impact on the automotive industry. A cumulative impact of $1.5-2.0 billion each month was noticed and evaluated across the industry. Despite phase wise unlocking and opening up, a steep decline in passenger vehicle demand played and is still playing a major role in the industry and its lack in exponential growth [24].

The Society of Indian Automobile Manufacturers (SIAM) said that overall automotive sales in the fiscal year that ended in March, India, the fifth-largest global market, hit a six year low (SIAM) that can be depicted using Figure 14.1. In 2019-20, a skeletal slowdown fueled by a slew of regulatory measures, as well as a stagnant economy, has placed vehicle sales on hold. In addition to the pandemic, which compounded sluggish sales [25]. In the midst of the rampant epidemic, restrictions, and lockdowns. For the third year in a row, the automobile sector is bracing for a difficult year. The overall auto industry's compound annual growth rate (CAGR) over the next five years (2015-16 to 2020-21, or FY21) is now negative at 2%, down from 5.7 percent in the previous five years (from 2010 - 16).The automotive industry's decadal growth has now fallen from 12.8 percent to 1.8 percent, demonstrating that there is more to the downturn than the pandemic, and that the epidemic solely cannot be cursed for multiple year lows in any segment in FY21 [26].

The shown below Figure 14.2 reveal the sales of the top two competitors of Maruti Suzuki in the Four-Wheeler industry and only Maruti Suzuki India Limited (MSIL) seems to have a positive magnification in terms of growth as compared to other homogeneous players out in the market. Sales in each segment literally approached multi-year lows in FY21, making it one of the industry's worst years ever. Passenger conveyance purchases in the domestic market fell to a six-year low with 2,711,457 units sold. In the domestic market, motorcycle and scooter purchases were also brushed off

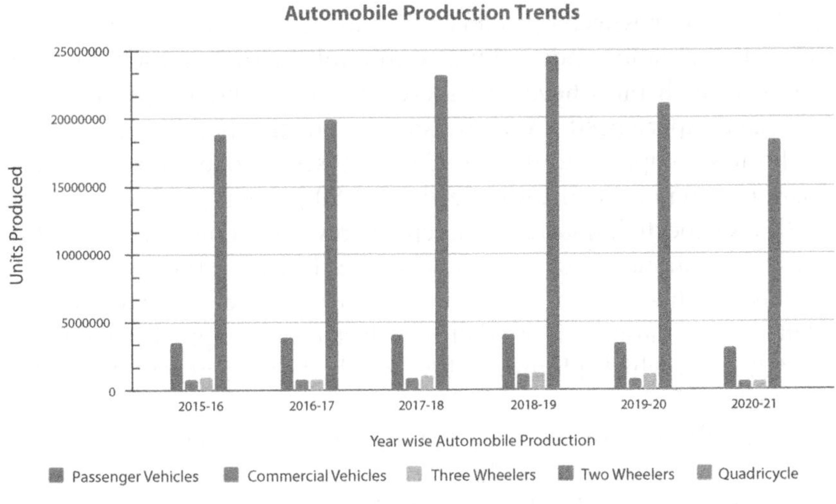

Figure 14.1 Automobile Production trends 2015–2021.

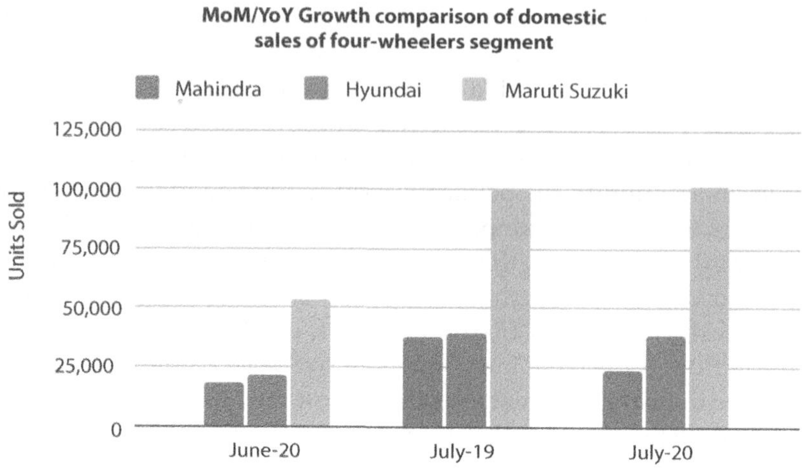

Figure 14.2 Domestic sales growth for four-wheelers segment.

to the 2014–2015 figures, with a volume of 15,119,000 units [18]. With 216,000 units sold, three-wheelers were the hardest hit, with volumes falling to a 19 years lowest sale. Furthermore, Commercial vehicle sales have also plummeted to their lowest point in over a decade.

14.2.2 Maruti Suzuki India Limited (MSIL) During COVID-19 and Other Players in the Automobile Industry and How MSIL Prevailed

India's largest four-wheeler producer, Maruti Suzuki, appears to be in command of the situation, not only have monthly sales increased, but year-over-year growth rates have also increased by around 1.3 percent. Sales figures have experienced a very substantial positive build-up in Monthly Growth rates compared to the pre-Covid-19 scenario, because practically all manufacturers have already reached 70-80 percent production capacity [27]. As expected, quarter-year reports revealed a bleak picture of the sector's whereabouts. Tata Motors, a market leader in the production of four-wheelers, had a poor "first quarter"—FY21. The scar that Covid-19 has left on the automobile market is reflected in compiled net-revenues and retail sales, which plummeted by nearly 48 percent and 42 percent, respectively [16].

In India, the size of the used automobile market/second-hand four-wheeler market is approximately 1.4 times that of new ones (in comparison to 4-5 times in the developed countries) and has a high magnification capacity [15]. Pre-COVID-19, Second-hand car sales were growing at a far quicker rate than new car sales, and industry insiders are already noticing an uptick in such sales. During the April-June period, it eventually led to used automobile online platform Droom to an increase of 175 percent in activity and a 250 percent increase in leads. From Figure 14.2, it can be concluded that during the same time period, Maruti Suzuki Veridical Value recorded a 15% increase in used car sales over the previous year. In June, Mahindra First Cull Wheels reported stronger demand than the previous year. Hyundai Motor India reported a magnification of 2% in domestic sales at 46, 866 units in August 2021 [4]. In the same period last year, the carmaker sold 45,809 vehicles, with sales hampered by the COVID and national restrictions on the import and export of components and implements.

14.3 Methodology

The methodology employed for this study to examine the impact of supplier networks on SCM of the Indian automotive sector and logistics management process post COVID-19 epidemic and how MSIL topped the Indian automotive list used a combination of literature review, single case study, and flexible methodology systematic approach [16] and the same is depicted using the flowchart in Figure 14.3. Secondary data was

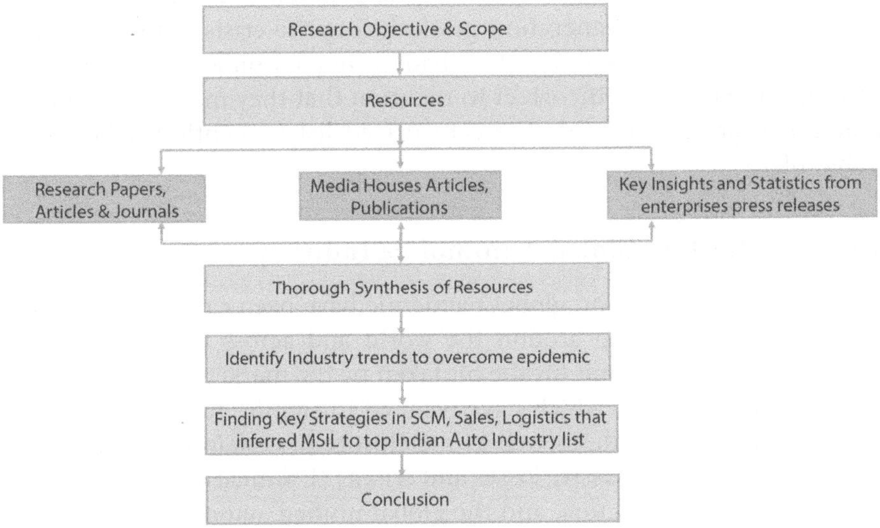

Figure 14.3 Flowchart of the research methodology.

used in the research, which included a literature study as well as printed media, social media and website articles. Information was gathered from research papers, news stories, related books, websites and company brochures [20].

The essay methodically arranged this material after a thorough examination of the same to capture the challenges and their solutions adopted by MSI, the chain of events that led to the case situation, and the steps made by MSIL to address the same in comparison to rivals' strategies. The data for this study came mostly from secondary sources, including the expert research and analytical articles, journals. Media coverage of the Indian auto industry, secondary data available in print and online/social media were also included [24]. SIAM Automotive Industry publications and annual reports and MSIL's own source via annual reports, press briefings, and other means.

14.4 Findings

14.4.1 Worldwide Economic Impact of the Epidemic

The effect of COVID-19 had a huge role to play in the direct sales and work flow of various industries especially in India. The impact of COVID-19 left a huge negative impact and will remain to be a big dent on our economy for the next years and generations to come as a prediction done by industrial experts and economists [27, 29]. Some sectors failed to hit a significant revenue generation mark during the crisis period whereas some of the industries and sectors had an impact on a very small scale or rather it will not be incorrect to mention that they made a significant amount of growth instead. The categorized list is mentioned below in Table 14.1.

14.4.2 Effect on Global Automobile Industry

The COVID-19 crisis and global pandemic have been causing disruption and economic hardship around the world and across the nations with boundaries being no limit for the hit taken by the market. No country has been spared its effects, which have resulted in significant economic stagnation and poor growth, as well as the closure of certain enterprises and organizations due to massive losses and crises [3]. Similarly, the disease has impacted other key sectors, and the global unified automobile sector has not been relinquished. The shutdown and forced closure of manufacturing

Table 14.1 Indian economy driving sectors Real Gross Value Added (GVA) growth comparison.

Real GVA Growth (in percentage)					
Sector	**2016-17**	**2017-18**	**2018-19**	**2019-20**	**2020-21**
I. Agriculture, Forestry and Fishing	6.8	6.6	2.6	4.3	3
II. Industry	8.4	6.1	5	−2	−7.4
II.i. Mining and Quarrying	9.8	−5.6	0.3	−2.5	−9.2
II.ii. Manufacturing	7.9	7.5	5.3	−2.4	−8.4
II.iii. Electricity, Gas, Water Supply and Other Utility	10	10.6	8	2.1	1.8
III. Services	8.1	6.2	7.1	6.4	−8.4
III.i. Construction	5.9	5.2	6.3	1	−10.3
III.ii. Trade, Hotels, Transport, Communication & Services related to Broadcasting	7.7	10.3	7.1	6.4	−18
III.iii. Financial, Real Estate and Professional Services	8.6	1.8	7.2	7.3	−1.4
III.iv. Public Administration, Defence and Other Services	9.3	8.3	7.4	8.3	−4.1
IV. GVA at Basic Prices	8	6.2	5.9	4.1	−6.5

companies, as well as the supply chain being impeded and disrupted as a result, decreased/lack of demand, have all taken their toll. As a result of their inability to cope with the losses, several auto dealers would close permanently, causing market share to plummet [6].

Car sales were one of the few businesses that existed prior to COVID-19, that had opposed the industry being shifted to the online platforms and converted it majorly to the E-Commerce market. The common pattern and research study have revealed that consumers browse out for vehicles over the internet and then visit the dealership stores to make the final purchase. So, the idea of it being shifted to complete online has been a major crack of an Idea and implementing it is under the works with major dealers having their own websites and online portfolios, it is a possibility that due to COVID and its impact online platforms and complete dealership being via online mode is not a lucid dream or imagination [14]. During the pandemic, surveys indicated that the percentage of customers who bought 50 percent or more of their total transactions online climbed from 25 percent to 80 percent, giving many businesses a chance to recoup their losses and weather the economic resurgence [28]. Despite the fact that recent market readings and figures showed hints of improvement month-over-month in August 2021, (MoM). The impact of this on many regions around the world has been discussed and defined as follows using the graph in Figure 14.4.

United States: The automotive industry in the United States is still in a precarious state. In August, sales dropped by nearly 20% (YoY). The shares of various major car brands being categorized below according to their

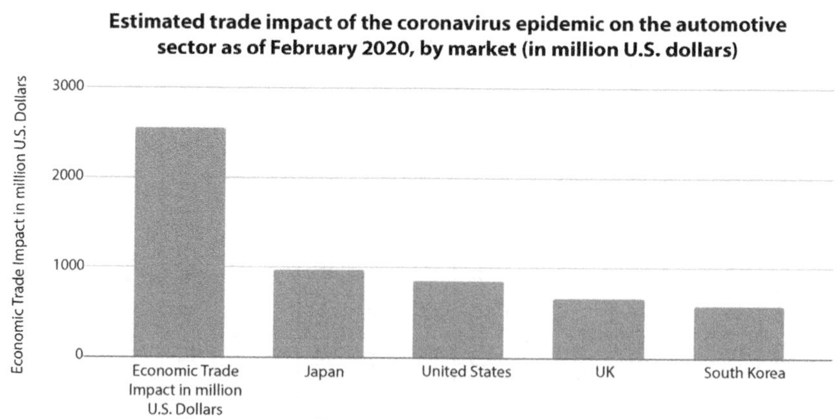

Figure 14.4 Global impact of COVID-19 on automotive sector.

stats are: Toyota suffered a 24.6 percent decrease, Honda at a net significant economic impact of (-23%). Hyundai, as compared to the other players out in the market performed significantly better with only an 8.4% decline overall [10].

European Union: In Europe, easing of lockdowns and recovery from COVID-19 has been better than other standout nations [18]. As a result, it surpassed 1.2 million manufactured items, down 16 percent on a year-over-year basis in comparison to previous year and is recovering at a better pace with each passing quarter and a far better improvement than others.

Japan: In Japan the chances of a speedier recovery and at a faster pace is suggested. Making it stand out to be better than the rest of the global competing nations. Following that, car sales increased by 11.6 percent year over year to 2.47 million units in H1 202 [15].

China: China's vehicle sales business continues to recover at a rapid pace. In August, vehicle shipments totaled close to 2.2 million units, up 11.6 percent year over year. Overall shipments during the January-August 2020 period were 10% lower on a year-over-year basis than they are now [28].

14.4.3 Effect on Indian Automobile Industry

The COVID-19 induced lockdown has had a major effect on the Automobile industry on a global basis, India as an economic zone has not been spared and has also faced a lot of shutdowns and closure of companies who could not survive the economic crisis surge. The same has also led to the disruption of the entire market chain system and the rotation of products as exports from India and auto parts as imports due to the shutdown of the whole nation on an emergency basis [27]. Adding to it the reduction in customer demand also had a huge role to play and it being the main source and contributory factor in the loss in revenue and severe liquidity crisis in the automobile sector. The other main reasons for the roadblock in the sales are as follows: the leapfrogging to BS6 emissions norms (effective from April 1 of 2020) from earlier BS4, constructive charges like GST. According to the studies and research done by the Society of Indian Automobile Manufacturers, the car industry in India alone witnessed a negative growth in sales of PVs (Passenger Vehicles) in FY21 a total of 2.24% decline as compared to earlier records, 13.19% fall in the sales of 2-wheelers, a hefty 20.77% negative growth in sales of CVs(Commercial vehicles) and an overall loss of 66.06% in sales of 3- wheelers [28]. Now coming up individually to the Auto Sector and its segmental analysis below, the stats show the comparison of sales and production of automobiles in FY 17'-20 in Figure 14.1. And the share of each segment in total production done in FY 2020 divided on the basis of vehicle types mainly prevailing in India, which can be inferred using Figure 14.5.

Maruti Suzuki India Limited (MSIL) cut down the temporary workforce by 6% due to the petty number of sales and drop in demand in the market. The auto sector which contributed around almost 7% of the nation's GDP is currently feeling the heat and is now facing a steep decline in the growth rate due to the COVID-19 scenario [24]. Along with MSIL the other players in the market altogether have observed a loss of more than 30% in recent months. Now, a recent study done in 2021 provided the facts on the analysis of the sales performance of the Auto Market participants and firms that is provided below in Table 14.2.

When compared to the same time the previous year, PV sales fell 17.88 percent in April-March FY 20'. In terms of PVs, sales of passenger cars and vans dipped by 23.58 percent and 39.23 percent, respectively, in April-March 2020, while sales of utility vehicles UVs ticked up by 0.48 percent [16]. The overall Commercial Vehicles segment fell by 28.75 percent in comparison to the same period last year, with Commercial Vehicles, Medium & Heavy Commercial Vehicles (M&HCVs), and Light Commercial Vehicles falling by 42.47 percent, 20.06 in FY 20' with record sales done during the same period in FY '19 which can be clearly seen using the above Table 14.2. [24] The sale of three-wheelers has decreased by 9.1 percent. In comparison to April-March 2019, passenger and goods carriers in the 3-Wheelers lost 8.28 percent and 13.27 percent, respectively in April-March 2020. In April-March 2020, the number of 2-wheelers decreased by 17.76 percent compared to the same period in 2019. Scooters and Motorcycles both lost 16.94 percent and 17.53 percent, respectively, in the 2-Wheelers market over the same time period [26].

MSIL leads the PVs segment and has a whooping share of 45.6% despite it being valued for being more than 50% in previous years. The next in the ladder was taken by Hyundai motors with 16.4% even though they saw a significant decline in their numbers for previous years. Third on the list being Tata motors with 9.3% in March and 8.8% in Feb 2021 another significant players Kia Motors, M&M, Toyota, Renault etc. being 6.0%, 5.2%, 4.7% ,3.9% respectively and other their MoM change during Feb'-Mar'21 is being represented in the tabular and graphical representations given below in Table 14.3 and Figure 14.6.

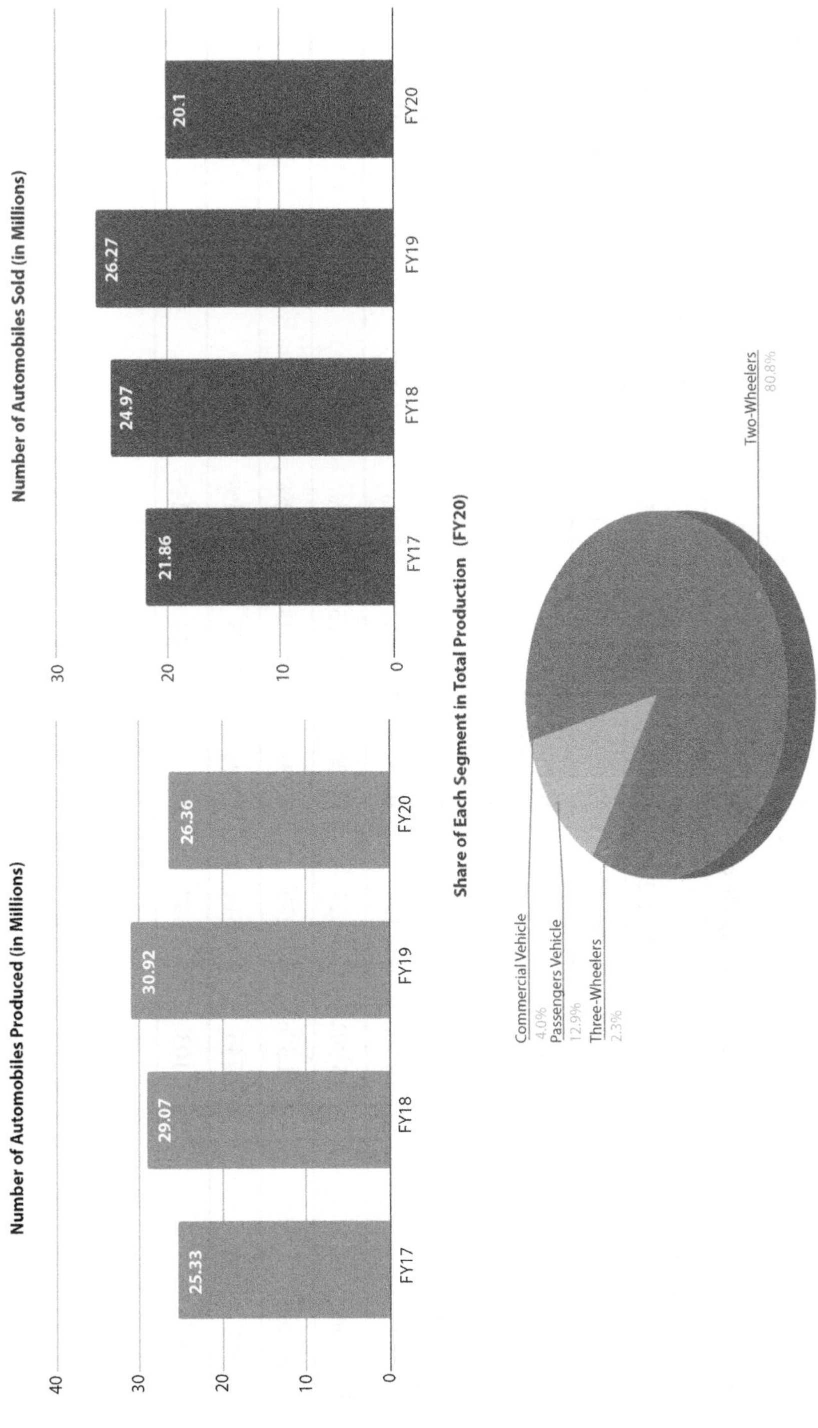

Figure 14.5 Sales percentage of vehicles according to their type.

Table 14.2 Stats during FY 19'-20' reflecting effect on sales.

PV Domestic Sales (Volume in Units)	Mar'21	Mar'20	YoY%	Feb'21	MoM%	FY21 (in Lakh)	FY20 (in Lakh)	YoY%
Maruti Suzuki	1,46,203	76,240	92%	1,44,761	1%	12.93	14.14	–8.50%
Hyundai Motors	52,600	26,300	100%	51,600	2%			
Tata Motors	29,654	5,676	422%	27,225	9%	2.22	1.31	69%
Kia Motors	19,100	8,583	123%	16,702	14%			
M&M	16,700	3,383	394%	15,391	9%	1.57	1.87	–16%
Toyota	15,001	7,023	114%	14,075	7%			
Renault	12,356	3,279	278%	11,043	12%			
Ford	7,746	3,519	120%	5,775	34%			
Honda	7,103	3,697	92%	9,324	–24%			

Table 14.3 Stats during Mar'21 and Feb'21 reflecting effect on sales.

Passengers Vehicle	Mar'21	Feb'21	MoM Change
Maruti Suzuki	45.60%	46.90%	–1.30%
Hyundai Motors	16.40%	16.70%	–0.30%
Tata Motors	9.30%	8.80%	0.40%
Kia Motors	6.00%	5.40%	0.60%
M&M	5.20%	5.00%	0.20%
Toyota	4.70%	4.60%	0.10%
Renault	3.90%	3.60%	0.30%
Ford	2.40%	1.90%	0.60%
Honda	2.20%	3.00%	–0.80%
MG	1.70%	1.40%	0.30%
Nissan	1.30%	1.40%	–0.10%
Volkswagen	0.63%	0.70%	–0.10%
Jeep	0.42%	0.40%	0.10%
Skoda	0.36%	0.30%	0.10%

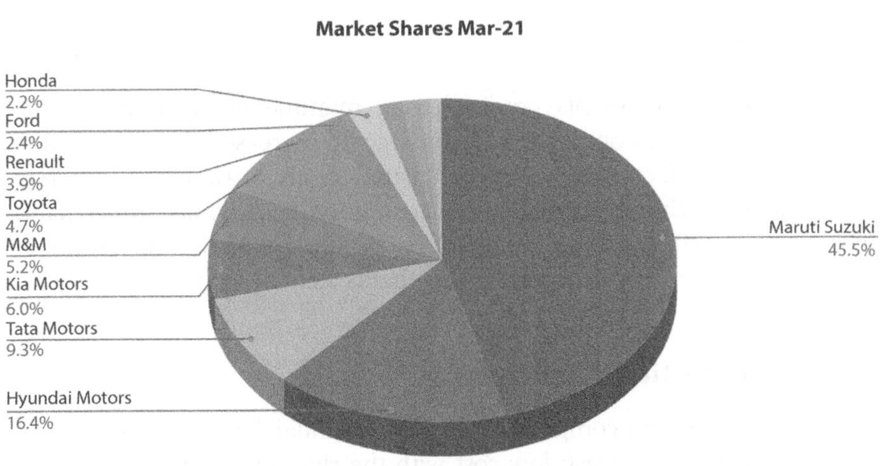

Figure 14.6 Market shares of different automotive sector players.

14.4.4 Automobile Industry Scenario That Can Be Expected Post COVID-19 Recovery

By the end of FY 2026, the $118 billion automobile market is expected to have grown to $300 billion. In FY 2020, India's year-out output was 26.36 million automobiles [27]. In FY20, the Pan-India automobile market, which includes two-wheelers and passenger vehicles, had an 80.8 percent and 12.9 percent net worth impact, respectively, resulting in a total turnover of nearly 20.1 million automobiles. In the coming years, passenger vehicles will dominate the market, closely followed by the mid-sized automobile industry. India's vehicle exports totaled 4.77 million units in FY20, representing a 6.94 percent CAGR from FY16 to FY20. Two-wheelers accounted for 73.9 percent of overall vehicle exports, with 14.2 percent going to passenger and mid-sized cars, 10.5 percent to three-wheelers, and 1.3 percent to commercial vehicles. Overall, we have witnessed a minor stabilization and recovery in majorly damaged industries, with automobiles being the only one where growth is entirely dependent on the individual success of multiple enterprises and market giants (across the country and via exports) [27].

Further in times to come, the auto industry can be boosted by government policies and decisions, such as reducing the basic cost of raw materials at a national level, implementing lower taxes and relaxation on taxes especially targeting the automobile sector. Steps like this can help the automobile industry recover at a much faster and stronger rate than expected and reach the global target researched by FY 2026 [25].

14.5 Discussion

There are a number of players in the car manufacturing industry who have given numerous options to customers and increased the competition among the manufacturers. Different customers are attracted by different values added to the product like low cost, good quality, fast and reliable delivery, availability, after-sale support etc. So understanding the customer requirements and providing the best in that is a challenging task [19].

14.5.1 Competitive Dimensions

MSIL's most significant competitors are Tata, Hyundai, Ford and Volkswagen. MSIL objective is to furnish low cost with the right quality product for the average income individual and as opposed to acquiring a broad segment it deals with niche and rule the industry. Following are the competitive dimensions it remains over others [8]:

- Cost: high customer satisfaction rating concerning the cost of ownership of Maruti Suzuki vehicles over the entirety of its range. MSIL concentrated on the niche market of compact cars offering useful features at a moderate cost.
- Quality: Maruti Suzuki car owners encounter fewer problems in their vehicle than some other car manufacturers in India. High quality is given inside at a reasonable price. In the premium compact car segment, Alto was chosen as number one.
- Delivery reliability and Speed: Maruti Suzuki has more than 307 state-of-the-art showrooms spread across 189 locations. Maruti Suzuki can provide faster service than its competitors in India because of its high localization.
- Flexibility and New product introduction speed: Maruti Suzuki has Japan based R&D. MSIL uses advanced innovation and technology to introduce models that fit in the current lifestyle with powerful engine efficiency. MSIL comes up with new variants in close intervals.
- Supplier after-sale product: this is one of the most significant advantages that Maruti has over others, cost of ownership, as well as the cost of maintenance, is very reasonable in case of Maruti Suzuki and high availability of its spare parts as well.

14.5.2 MSIL Strategies

MSIL's significant strategies for maintaining its position atop the Indian car market segment during multiple downturns include:

In 1991, the first phase of liberalization was declared, and automobile segments were permitted to have foreign collaboration. The Government of India teamed up with Suzuki Inc. (Japan) to create India's most popular car, the "Maruti." Suzuki helped Maruti component makers to overhaul their technology and appropriation of Japanese benchmarks of quality. The Indian passenger automobile market was driven and guided up to a maximum proportion of it from that point forward [19].

The competition raised with the new entries of global car makers and analyzing heat of competition from global carmakers, MSIL implemented an extensive strategy for accruing and retaining the customers. The strategy was to provide finance and insurance of cars and sell out or purchase a pre-owned car and this led MSIL into another business and carried huge customers and additional revenue to MSIL thus expanding its network and pulling customers for MSIL [8].

Maruti always attempts to reduce the cost and reinforce the quality throughout its value chain, which led to substantial progress of MSIL. The company propelled five vehicles in CNG variants in a day (Estilo, Alto, SX4, WagonR, and Eeco).In Manesar, MSIL established two new Greenfield production lines, which boosted production and allowed the company to produce 1.85 million units by the end of 2012 [19].

MSIL aims to strengthen its network in rural sector dealers and suppliers to make a firm grasp over the rustic market. The aim is to get more and more local vendors to reduce logistic cost, crude material cost and maintain JIT and diminish inventory cost also. MSIL focuses on providing new models, fuel-efficient and cost-effective products that do not squeeze the customer pocket much and furthermore satisfy their aspirants. So customer satisfaction at least expense is their definitive objective [22]. As a result, Maruti Suzuki had to maintain its quality while delivering a less expensive vehicle. Also, if it imports components from Japan, it would be reasonably expensive; thus, it started putting efforts in developing its domestic component makers while not only reducing the cost but also increasing the availability.

MSIL stepped forward to make firm and cohesive suppliers and dealer's network by hedging bank financing for them. By assisting their suppliers, MSIL strengthened its hold over them, getting additional values and more favorable conditions for the company for present and future deals. MSIL ruled the car manufacturing industry in small car segments with its two most profitable products, Maruti 800 and Alto. This segment of car manufacturing is becoming profoundly competitive, with quickly expanding the number of players coming up with new models. One of the most appropriate examples is Tata –Nano which competes with Maruti 800 and brings down its share. MSIL decided for contraction defense strategy wherein it ceased the Maruti 800 production and left Nano to move over the lower segment of the car market [21].

14.5.3 MSIL Operations and Supply Chain Management

Comprehensively Supply chain the management (SCM) can be described as the way toward planning, executing, implementing, tracking and controlling the tasks that go into improving how an organization buys the crude segments it needs to make products or services further fabricates or manufactures those products or services and lastly supply it to the customers in the most effective way possible [27]. A supply chain includes all parties involved in satisfying a consumer request, whether directly or indirectly, such as transporters, warehouses, retailers, and customers themselves. A supply chain is a dynamic system that incorporates the continuous flow of information, products, and assets between phases. Operational information related to the production process ought to be shared among manufacturers and suppliers so as to make supply chains effective. The ultimate goal is to build, establish, and coordinate the production process across the supply chain in such a way that the competition will struggle to find a match. MSIL is one of the most prominent and greatest supply chain and logistics management tales in the automotive industry. Throughout the years, it has worked hard to transform problems into possibilities and obstacles into opportunities [4].

14.5.4 MSIL Suppliers Network

Ten percent of the components in Maruti production is directly sourced from foreign markets and its local vendor's imports another 10% to 15% of the components. There are 800 local suppliers, including Tier I, Tier II, and Tier III providers, as well as 20 foreign suppliers who work together in a consistent manner. MSIL intends to reduce its disclosure to the foreign trade by half over the next few years in order to reduce turbulence. Maruti's domestic Tier I is leaner, with only 246 suppliers, 19 of which it has formed joint ventures with and maintains significant equity stakes in to maintain a state of production and quality [21].

Top management of MSIL have analyzed that one the significant thing for prevailing upon challenges in this competitive market scenario is to have vast and cohesive suppliers or vendors network and therefore from the beginning, MSIL attempts to improve conditions at vendors end as follows [18]:

- Localization of suppliers and components: To avoid the fluctuation in currency shifts and high cost of logistics, localization is one of the significant mantras of Maruti Suzuki's supply chain development over the previous decade.
- Huge supplier base: They cooperate with a large number of suppliers just as they manage suppliers to accomplish profound cost decrease year on year. Also, if it imports components from Japan, it would be reasonably expensive; thus, it started putting efforts in developing its domestic component makers while not only reducing the cost but also increasing the availability.

- Massive investment in suppliers: some measures are designed and implemented to help and support suppliers. Maruti gets authorization from India's central bank to hedge currency for the benefit of Indian suppliers. The carmaker also acquires crude material in bulk for suppliers and orchestrates low-cost borrowing for suppliers to help companies obtain a better deal. Payments are also designed for a little cost, with only a nine-day cycle from the date of invoice accommodation.
- Shared savings programs: a mutual savings program has been presented by Maruti for its suppliers, called "value analysis value engineering." This program says rather than importing crude, suppliers should localize it and saving benefits will be shared among all.

14.5.5 MSIL Manufacturing

- Maruti Suzuki was tasked with creating a "people's automobile" that was both affordable and of high quality. Maruti Suzuki's first move was to create a high production standard. MSIL's plan for lowering production costs and improving quality was to use economies of scale.
- Phased Manufacturing Program: PMP ordered foreign firms to promote localization. MSIL deals with 50% local suppliers in the first 3 years then 70% by 5th year. MSIL's early focus was on the local market rather than export, which allowed it to negotiate less on the quality of components provided by producers, something it could not do if it were exporting.
- Location of supplier: Conventional automobile industry was in Tamil Nadu, and Maharashtra Maruti Suzuki has its manufacturing plant away from them, which makes transportation very inefficient. Thus for a better supply of material, it was required to locate the suppliers, and component makers close to the manufacturing plant of MSIL and JIT system added more necessity to it. In this manner, MSIL persuaded its suppliers from various Indian states to locate their manufacturing facilities near MSILs.
- Just In Time (JIT): MSIL was the first automaker in India to implement the JIT technique. The JIT system demanded that all manufacturers and suppliers be adequately trained

to meet the manufacturer's needs in a timely manner [20]. Furthermore, for quick, reliable and on-time delivery of material MSIL has localized its suppliers nearby manufacturing plant. This likewise diminishes detailed on-site inspections and testing of material done by MSIL.

- Lean manufacturing: Maruti Production System (MPS) uses lean manufacturing where they accelerate the speed of manufacturing and lower the cost, add more value to the customer so that customer would be willing to pay for it and reduce the waste by doing the right thing at the first time and eliminating the things that do not add value to the customer or less required. In lean production, they use JIT, Pull system/Kanban, continuous flow of work, and eliminate wastes: overproduction, excessive inventory, underutilization of workforce, waiting. They use the Kaizen improvement method for employees. They try to build quality into their process, which saves additional audits later, and also uses mistake-proofing [21].

14.5.5 MSIL Distributors Network

Previously, buyers would place an order for a vehicle and wait for over a year to receive it. Furthermore, the concept of Showrooms was non-existent, and the state of after-sales support was far worse. Maruti stepped up with the purpose of changing this situation and providing better client service. Maruti Suzuki built up a distinctive distribution network for gaining the competitive advantage. The company currently has 802 sales centers in 555 towns and cities, as well as 2740 customer support workshops in 1335 towns and cities. The primary goal of establishing such a vast distribution network was to reach out to clients in remote places and deliver the company's products. MSIL utilized the following techniques to boost dealer competitiveness and hence their profit margins.

The corporation would occasionally give out special awards for certain categories of sales. Various Opportunities were offered to dealers by Maruti Suzuki to earn more profits by different avenues given by MSIL like pre-owned car sales and purchase or finance and insurance services. MSIL established 255 customer service facilities in 2001-02 in combination with 21 highway segments, dubbed the Non Stop Maruti Express Highway. Out of 15,000 dealer sales executives, 2,500 were rural dealer sales executives in the year 2008 in MSIL [15].

14.5.6 MSIL Logistics Management

Since more than 30 per cent in transportation there are logistics costs, therefore operating productively and efficiently makes great financial sense [15]. Additionally, there is a crucial role in customer service levels as well as a geographic location in this plant set up decisions. For efficacious management of transportation Shipment sizes and routing and scheduling of equipment are one of the most important things to be considered. For better coordination and logistics management, sensitive demand and sales data, updated to date inventory data, stock and all shipment status information must flow on time between whole networks [17]. Transparency in Supply chain networks increases the visibility and adaptability characterized by the SCM can frequently lead to effective logistics management. In 1992 the lead time of MSIL was 57 days, but it has reduced to 19 days in 2013 and has diminished all the more at present [18].

14.6 Conclusion

The Indian Automobile market today is very dynamic & has been very competitive and will additionally get more with a scope of more players and products to enter. In the present vicious rivalry, it is extremely hard to endure. In India, MSIL is leading Automaker Company which possesses eminent position because of its extensive local supplier network and its implemented strategies in supply chain and logistics management improves the efficiency and performance of the entire value chain while also providing numerous benefits to all value chain partners in terms of lowering inventory and transportation costs, achieving lean operations and shorter time to manufacture a product, integrating valuable partners, and increasing product availability.

References

1. Alaadin, M., Covid-19: The Impact on the Manufacturing Industry. *Marsh*, 202019, https://www.marsh.com/content/dam/marsh/Documents/PDF/MENA/energy_and_power_industry_survey_results.pdf.
2. Alam, M.N., Alam, M.S., Chavali, K., Stock market response during COVID-19 lockdown period in India: An event study. *J. Asian Finance, Econ. Bus.*, *7*, 7, 131–137, 2020, doi: https://doi.org/10.13106/jafeb.2020.vol7.no7.131.
3. Belhadi, A., Kamble, S., Jabbour, C.J.C., Gunasekaran, A., Ndubisi, N.O., Venkatesh, M., Manufacturing and service supply chain resilience to the

COVID-19 outbreak: Lessons learned from the automobile and airline industries. *Technol. Forecast. Soc Change*, *163*, 120447, 2021 October 2020, doi: https://doi.org/10.1016/j.techfore.2020.120447.

4. Bhatt, P. and Varghese, S., Strategizing under economic uncertainties: Lessons from the COVID-19 pandemic for the Indian auto sector. *J. Oper. Strateg. Plan.*, *3*, 2, 194–225, 2020, doi: https://doi.org/10.1177/2516600x20967813.
5. Bhattacharya, S., Supply chain management in Indian automotive industry: complexities, Challenges and way ahead. *Int. J. Manage. Value Supply Chain.*, *5*, 2, 49–62, 2014, doi: https://doi.org/10.5121/ijmvsc.2014.5206.
6. Breja, S.K., Banwet, D.K., Iyer, K.C., Quality strategy for transformation: A case study. *TQM J.*, *23*, 1, 5–20, 2011, doi: https://doi.org/10.1108/17542731111097452.
7. Cai, M. and Luo, J., Influence of COVID-19 on manufacturing industry and corresponding countermeasures from supply chain perspective. *J. Shanghai Jiaotong Univ.,Sci*, *25*, 4, 409–416, 2020, doi: https://doi.org/10.1007/s12204-020-2206-z.
8. Corporation, S. M. (n.d.), *Supplier development in Indian auto industry: Case of maruti suzuki india limited.* https://core.ac.uk/download/pdf/230430874.pdf.
9. Frohlich, M.T. and Westbrook, R., Arcs of integration: An international study of supply chain strategies. *J. Oper. Manage.*, *19*, 2, 185–200, 2001, doi: https://doi.org/10.1016/S0272-6963(00)00055-3.
10. Ishida, S., Supply chain management in Indian automotive industry :complexities, Challenges and way ahead. *IEEE Eng. Manage. Rev.*, *48*, 3, 146–152, 2020, doi: https://doi.org/10.1109/EMR.2020.3016350.
11. Jha, H.M., Srivastava, A.K., Bokad, P.V., Deshmukh, L.B., Mishra, S.M., Countering disruptive innovation strategy in Indian passenger car industry: A case of Maruti Suzuki India Limited. *South Asian J. Bus. Manag. Cases*, *3*, 2, 119–128, 2014.
12. Julka, T., Administration, B., College, S. S. J. S. P. G., Suzuki, M., Supply chain and logistics management innovations at Maruti Suzuki India Limited. *Int. J. Manage. Soc. Sci. Res.*, *3*, 3, 41–46, 2014.
13. Krishnaveni, M. and Vidya, R., *Growth of indian automobile industry*, International Journal of Current Research and Academic Review-(IJCRAR) ISSN - 2347 - 3215, vol. *3*, pp. 110–118, 2015.
14. Kumar, R., Singh, R.K., Shankar, R., Study on coordination issues for flexibility in supply chain of SMEs: A case study. *Glob. J. Flex. Syst. Manage.*, *14*, 2, 81–92, 2013, doi: https://doi.org/10.1007/s40171-013-0032-y.
15. Kumar, V. and Gautam, V., Maruti Suzuki India Limited: The celerio. *Emerald Emerg. Mark. Case Stud.*, *5*, 1, 1–13, 2015, doi: https://doi.org/10.1108/EEMCS-03-2014-0058.
16. Lokhande, M.A. and Rana, V.S., Marketing strategies of Indian automobile companies: A case study of Maruti Suzuki India Limited. *SSRN Electron. J.*, *1*, 2, 40–45, 2016, doi: https://doi.org/10.2139/ssrn.2719399.
17. Nayak, J., Mishra, M., Naik, B., Swapnarekha, H., Cengiz, K., Shanmuganathan, V., An impact study of COVID-19 on six different industries: Automobile,

energy and power, agriculture, education, travel and tourism and consumer electronics, in: *Expert systems*, 2021.

18. Okorie, O., Subramoniam, R., Charnley, F., Patsavellas, J., Widdifield, D., Salonitis, K., Manufacturing in the time of COVID-19: An assessment of barriers and enablers. *IEEE Eng. Manage. Rev.*, *48*, 3, 167–175, 2020, doi: https://doi.org/10.1109/EMR.2020.3012112.
19. Paul, S.K. and Chowdhury, P., A production recovery plan in manufacturing supply chains for a high-demand item during COVID-19. *Int. J. Phys. Distrib. Logist. Manage.*, *51*, 2, 104–125, 2021, doi: https://doi.org/10.1108/IJPDLM-04-2020-0127.
20. R., R., Flexible business strategies to enhance resilience in manufacturing supply chains: An empirical study. *J. Manuf. Syst.*, *60*, October 2020, 903–919, 2021, doi: https://doi.org/10.1016/j.jmsy.2020.10.010.
21. Kiran Raj KM, Nandha Kumar KG -Impact of Covid-19 pandemic in the automobile industry: A case study, I*nternational Journal of Case Studies in Business, IT and Education (IJCSBE)*, Volume 5 Issue 1 Pages 36-49, 2021
22. Sahoo, T., Banwet, D.K., Momaya, K., Strategic technology management in the auto component industry in India: A case study of select organizations. *J. Adv. Manage. Res.*, *8*, 1, 9–29, 2011, doi: https://doi.org/10.1108/09727981111129282.
23. Shah, M.K. and Tomer, S., How brands in India connected with the audience amid Covid-19. *Int. J. Sci. Res. Publ.*, *10*, 8, 91–95, 2020, doi: https://doi.org/10.29322/ijsrp.10.08.2020.p10414.
24. Since january 2020 Elsevier has created a COVID-19 resource centre with free information in English and Mandarin on the novel coronavirus COVID-19, in: *The COVID-19 resource centre is hosted on elsevier Connect , the company's public news and information . (2020)*, January, 2020–2022.
25. Singh, N. and Salwan, P., Contribution of Parent company in growth of its subsidiary in emerging markets: Case study of Maruti Suzuki. *J. Appl. Bus. Econ.*, *17*, 1, 24, 2015.
26. Singh, T., Challenges in automobile industry in india in the aftermath of Covid-19, *17*, 6, 6168–6177, 2020.
27. Wu, X., Zhang, C., Du, W., An analysis on the crisis of "chips shortage" in automobile industry - Based on the double influence of COVID-19 and trade friction. *Journal of Physics: Conference Series*, vol. *1971*, 2021, doi: https://doi.org/10.1088/1742-6596/1971/1/012100.
28. Xu, Z., Elomri, A., Kerbache, L., El Omri, A., Impacts of COVID-19 on global supply chains: Facts and perspectives. *IEEE Eng. Manage. Rev.*, *48*, 3, 153–166, 2020, doi: https://doi.org/10.1109/EMR.2020.3018420.
29. Swetha, K.R. and N. M, A. M. P and M. Y. M, Prediction of pneumonia using big data, deep learning and machine learning techniques. *2021 6th International Conference on Communication and Electronics Systems (ICCES)*, pp. 1697–1700, 2021, doi: 10.1109/ICCES51350.2021.9489188.

About the Editors

M. Niranjanamurthy, PhD, is an assistant professor in the Department of Computer Applications, M S Ramaiah Institute of Technology, Bangalore, Karnataka. He earned his PhD in computer science at JJTU, Rajasthan, India. He has over 11 years of teaching experience and two years of industry experience as a software engineer. He has published several books, and he is working on numerous books for Scrivener Publishing. He has published over 60 papers for scholarly journals and conferences, and he is working as a reviewer in 22 scientific journals. He also has numerous awards to his credit.

Kavita Sheoran, PhD, she is an associate professor in the Computer Science Department, MSIT, Delhi, and she earned her PhD in computer science from Gautam Buddha University, Greater Noida. With over 17 years of teaching experience, she has published various papers in reputed journals and has published two books.

Geetika Dhand, PhD, is an associate professor in the Department of Computer Science and Engineering at Maharaja Surajmal Institute of Technology. After earning her PhD in computer science from Manav Rachna International Institute of Research and Studies, Faridabad, she has taught for over 17 years. She has published one book and a number of papers in technical journals.

Prabhjot Kaur, has over 19 years of teaching experience and has earned two PhDs for her work in two different research areas. She has authored two books and more than 40 research papers in reputed journals and conferences. She also has one patent to her credit.

Index

Also of Interest

By the same editors

ADVANCES IN DATA SCIENCE AND ANALYTICS, Edited by M. Niranjanamurthy, Hemant Kumar Gianey, and Amir H. Gandomi, ISBN: 9781119791881. Presenting the concepts and advances of data science and analytics, this volume, written and edited by a global team of experts, also goes into the practical applications that can be utilized across multiple disciplines and industries, for both the engineer and the student, focusing on machining learning, big data, business intelligence, and analytics.

WIRELESS COMMUNICATION SECURITY: Mobile and Network Security Protocols, Edited by Manju Khari, Manisha Bharti, and M. Niranjanamurthy, ISBN: 9781119777144. Presenting the concepts and advances of wireless communication security, this volume, written and edited by a global team of experts, also goes into the practical applications for the engineer, student, and other industry professionals.

MEDICAL IMAGING, Edited by H. S. Sanjay, and M. Niranjanamurthy ISBN: 9781119785392. Written and edited by a team of experts in the field, this is the most comprehensive and up-to-date study of and reference for the practical applications of medical imaging for engineers, scientists, students, and medical professionals.

SECURITY ISSUES AND PRIVACY CONCERNS IN INDUSTRY 4.0 APPLICATIONS, Edited by Shibin David, R. S. Anand, V. Jeyakrishnan, and M. Niranjanamurthy, ISBN: 9781119775621. Written and edited by a team of international experts, this is the most comprehensive and up-to-date coverage of the security and privacy issues surrounding Industry 4.0 applications, a must-have for any library.

Check out these other related titles from Scrivener Publishing

CONVERGENCE OF DEEP LEARNING IN CYBER-IOT SYSTEMS AND SECURITY, Edited by Rajdeep Chakraborty, Anupam Ghosh, Jyotsna Kumar Mandal and S. Balamurugan, ISBN: 9781119857211. In-depth analysis of Deep Learning-based cyber-IoT systems and security which will be the industry leader for the next ten years.

MACHINE INTELLIGENCE, BIG DATA ANALYTICS, AND IOT IN IMAGE PROCESSING: Practical Applications, Edited by Ashok Kumar, Megha Bhushan, José A. Galindo, Lalit Garg and Yu-Chen Hu, ISBN: 9781119865049. Discusses both theoretical and practical aspects of how to harness advanced technologies to develop practical applications such as drone-based surveillance, smart transportation, healthcare, farming solutions, and robotics used in automation.

MACHINE LEARNING TECHNIQUES AND ANALYTICS FOR CLOUD SECURITY, Edited by Rajdeep Chakraborty, Anupam Ghosh and Jyotsna Kumar Mandal, ISBN: 9781119762256. This book covers new methods, surveys, case studies, and policy with almost all machine learning techniques and analytics for cloud security solutions.

Printed and bound by CPI Group (UK) Ltd, Croydon, CR0 4YY
20/08/2023
08102001-0005